AF869064

The

FRIENDLY

IONOSPHERE

Also from *TYNDAR PRESS*

Sparks and Flames
Ignition in Engines: An Historical Approach
Crawford MacKeand
LCN 97-90492 ISBN 0-9659066-4-7

The FRIENDLY IONOSPHERE

by

Crawford MacKeand

WA3ZKZ/VP8CMY

Signals, Noise and Propagation

TYNDAR PRESS

Published by *Tyndar Press*,
P.O. Box 236, Montchanin DE 19710 U.S.A.

Copyright © by J. Crawford B. MacKeand 2001
All rights reserved

No part of this book may be reproduced or transmitted in any form, or by any means, mechanical or photographic or electronic, or by an information storage or retrieval system, without the publisher's written permission, except for the customary use of quotations in a review.

Cataloging-in-Publication Data

MacKeand, Crawford

The Friendly Ionosphere: Signals, Noise and Propagation
/ J. Crawford B. MacKeand

Bibliography p. 112 - 115 and index
1. Radio wave propagation
2. Electromagnetic interference
3. Radio noise
4. Radio - interference
5. Signal processing
I. Title
TK 6553.M33 2001 621.384 11 dc-20
ISBN 0-9659066-5-5

Library of Congress Control Number: 2001126286

Printed and bound in the United States of America.

FOREWORD

The propagation of H.F. radio waves is one piece of a larger puzzle - how do radio signals pass from one station to another, and how well do they pass? Can the ionosphere support the desired propagation; is the proposed operation above or below the Maximum Usable Frequency or MUF? And is the ionosphere in fact required at all? But these yes or no answers to propagation must be combined with many other pieces of the puzzle. The strength of the transmitter is usually fairly well understood, but the gain of the antennas at the angles at which the signals actually leave and approach is less well known. Antenna efficiency is often an almost closed book to us. In the propagation path itself, there are many possible ways to lose the signal, spreading and D region absorption usually being most important, but losses on reflection and those factors often lumped together as Over-MUF losses must be considered too.

However, in the end, none of this will be of much value if the incoming signals cannot be usefully distinguished from radio noise at the receiver. So we must also know the probable noise level, no matter whether it's from atmospheric, man-made or even cosmic sources. Also, many different modulation schemes carry our messages on H.F. radio waves, and they vary greatly in how much or how little they are affected by the often strenuous competition which noise offers.

Much has been written on the subject, but I have tried to bring it together so that it is clearly understandable to anyone with a scientific interest in H.F. operation and its possibilities. I hope also that the case studies in the final chapters will not only illustrate the problems, but tell their own intrinsically interesting stories.

Crawford MacKeand, WA3ZKZ, VP8CMY
Greenville, Delaware, U.S.A.
March 2001

CONTENTS

ABBREVIATIONS

AGW	Atmospheric Gravity Wave
AM	Amplitude Modulation
ARRL	American Radio Relay League
CATV	Cable Television
CCIR	Comité Consultatif Internationale (Radio), now ITU
CRPL	Central Radio Propagation Laboratory
CW	Continuous Wave (usually International Morse Code)
DSB	Double Sideband
EHV	Extra High Voltage
EM	Electro-magnetic
FM	Frequency Modulation
FOT	Frequence Optime de Travail (Optimum Working Frequency)
HF	High Frequency (3 - 30 MHz)
IEE	Institution of Electrical Engineers (London)
IEEE	Institute of Electrical and Electronics Engineers (NY)
ITU	International Telecommunication Union
JTAC	Joint Technical Advisory Committee (IEEE)
KHAQQ	1937 Earhart Aircraft Callsign
LUF	Lowest Usable Frequency
MF	Medium Frequency (0.5 - 3 MHz)
MUF	Maximum Usable Frequency
NBS	National Bureau of Standards
NVIS	Near Vertical Incidence Service
PCA	Polar Cap Absorption
PSK	Phase-Shift Keying
RCA	Radio Corporation of America
RMS	Root Mean Square
RTTY	Radio Teleprinter
RTW	Round-the-World
SID	Sudden Ionospheric Disturbance
SSB	Single Sideband
SWL	Short-Wave Listener
TID	Traveling Ionospheric Disturbance
TOR	Teleprinter on Radio
VHF	Very High Frequency (> 30 MHz)
WE	Western Electric
WWV	Standard Frequency Station (USA)

CHAPTER I

INTRODUCTION

The progress of short wave radio signals from transmitter to receiver has been studied ever since long distance high frequency (HF) communication was first established in the 1920s, and an enormous body of information exists on this fascinating subject. Unfortunately for non-specialists, especially amateur non-specialists, much of it is hard to apply; this has been a happy hunting ground for mathematically minded physicists. However, a different view can consider the engineer's *"How?"* rather than the scientist's *"Why?"*, and so, after a quick glance at *Why*, we examine the *Hows* of HF ionospheric communication, with a brief nod to forward scatter, ground-wave and other interesting modes. The HF frequency spectrum is usually defined as 3 - 30 MHz, but we will extend our range slightly into the medium frequencies (MF) from 1.5 - 3 MHz.

High frequency ionospheric sky-wave communications rely on conductive zones of ionized gas, mainly two known as the E and F regions, about 100 and 400 Km above us in the earth's atmosphere. Their ionization varies with season, time of day and solar activity, in part described by solar flux or sunspot number. These ionized zones deflect and return some radio signals to listeners below, while other signals go on into outer space, lost to earthly communication. Although the zones are not really layers, and the process is not properly a reflection, the terms are so well-established, convenient and familiar that we will use them nonetheless. Useful sky-wave radio routes must intersect layers which support communication, and if they do reflect the signal, then not too much of it must be lost *en route*. Long distances may be covered by multiple reflections, ionosphere to earth or to ocean surface and back again. The largest share of end-to-end loss or attenuation is due simply to the signal spreading out in space, but we will also want to know how much the signals are absorbed passing through and reflecting from the layers up above, or as they repeatedly reflect from earth and ocean on longer routes. Absorption loss depends especially on the ionization of the D layer at 60 - 90 Km and on the working frequency. Auroras often affect transpolar paths; on trans-equatorial routes layer shape and density over the equator may be important; antipodean paths offer focussing. Clearly, signal strength into the receiver depends on the antennas at both ends of the circuit, but it may be less clear that antenna gain is very dependent on the elevation angle at which signals leave or approach our stations, and on the terrain near the antennas. Finally, other minor losses such as antenna feeder attenuation must also be factored into any complete overview.

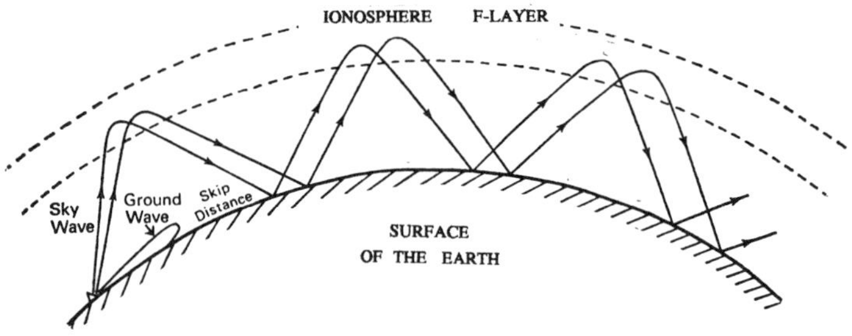

Figure 1.1 Ionospheric signal paths.

No signal, no matter how strong, has any value to us if we can't disentangle its information or modulation from the inevitable accompanying noise. We therefore need to find out how loud the noise will be and how well the modulation system will tolerate that noise; also whether our station is set up to reject unwanted noise, by antenna directivity for example. In general, noise also includes interference, and today in our high frequency bands, interference from other stations is usually by far the most difficult problem. It is harder to quantify and to the extent that it is an operational rather than technical problem, it will fall outside our scope. In evaluating signals against atmospheric noise our principal data source will be the International Telecommunication Union reports, with their comprehensive world-wide noise grade maps. Cosmic or galactic and man-made noise must be added in, and also internal receiver noise, normally important only at VHF and above.

Sky-wave signals for any terrestrial destination must find an ionosphere able to bend them to the desired path. The bending process is such that we can only operate below a maximum take-off angle and below a maximum usable frequency or MUF. Higher angle rays pass straight into outer space, usually leaving a "skip zone" close to the transmitter in which no sky wave is received. But, MUF is only a statistical value. The classical definition says, roughly, that a route will be workable up to a little above the MUF on half the days of a month, and up to a little below MUF on the other days, the average highest workable frequency being the MUF. Many HF propagation books treat MUF calculations in depth, but here, after we launch the signal on its hopeful way from the transmitter in Chapter III, it will only get a "once over lightly" treatment in Chapter V. Finally, not all signal paths are ionospheric. Beyond the skip zone, signals are returned from the ionosphere, but between the transmitter and the near edge of the skip zone (if indeed one exists), ground-wave propagation holds, with rules quite different from the sky-wave. Energy loss to the earth's surface ties ground-wave performance very tightly to the conductivity of that surface between transmitter and receiver.

So, first we must define the geographic path, and then establish that transmission is at least possible. The following several chapters describe the many and various hindrances that signals meet between transmitter output and receiving site, and also how the remaining signal's strength can be determined as each hindrance is passed. Then we will take a hard look at the competitive entry, noise, and its strengths and weaknesses, and finally we will investigate how well our receiving station set-up can handle all the competition.

CHAPTER II

THE PATH FROM TRANSMITTER TO RECEIVER

Sky-wave signals are very dependent on the local conditions that they encounter *en route* from transmitter to receiver. Therefore, we will have to define in some detail the path which the signal energy will take from beginning to end. The presence, absence or density of the necessary ionospheric reflecting layers depends on local time and the sun's zenith angle at the point of reflection, as well as on other solar effects. Likewise many losses; for example, the basic absorption loss in the ionosphere will vary from North to South, and with local time of day.

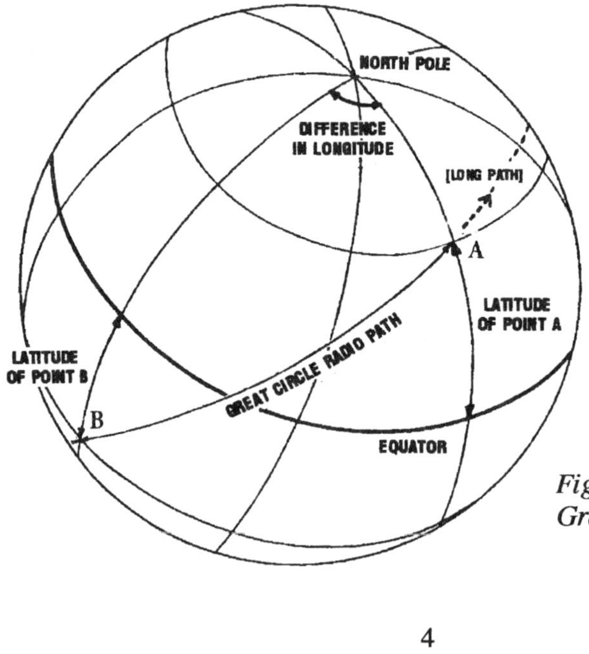

Figure 2.1 Stations and Great circle path.

THE PATH FROM TRANSMITTER TO RECEIVER

Radio path calculation is very straightforward mathematically, a geometrical rather than a radio problem, and we will therefore review it in outline only. Radio waves use the most direct available routes; in principle, great circles around the world. Great circle trigonometry is merely tedious to calculate, with a little complexity in getting the signs right on the longer paths! Only five data items are required, the latitude and longitude of the transmitting and receiving stations, and whether short or long path is wanted. From our input and its manipulation, and certain reference information, converting geographical co-ordinates to magnetic equivalents for instance, we will finally want to see a number of position co-ordinates on the path, geographic and magnetic, and also antenna bearing angles.

However, there's a more multi-dimensional aspect. We can't just define a linear route over the sphere's surface, but must also identify the number of hops the signal must make between earth and ionosphere, and the reflecting site locations. We may also identify the route's mid-point, as one simple method of calculating all these points is to divide the path in two and determine distances and positions along the route from the ends into the middle. Then to switch from short path to long we just exchange mid-point latitude and longitude for those of its antipode.

But we are ahead of ourselves. The length of each hop is also set by the height of the ionosphere, and since the height at F layer reflection points is usually obtained from a set of world maps, the co-ordinates of each point are required first. And of course we don't yet know whether either layer will be where we need it! To solve this dilemma, we can start with a rough estimate, and then calculate final locations and heights in just a few iterations. (The E layer is easier: its height can reasonably be taken as constant.) We can now determine ray angles for the antennas at both ends, and incidence angles at the intermediate points too. Estimates of terrestrial reflection loss also depend on knowing geographical locations, as the losses at earth points are rather higher than at oceanic points, while both of them depend on angles of incidence. And, as we will see later, to determine auroral losses we rely on knowing the intersections of the path and the North or South polar auroral ovals, and the local times for those intersections. But now that the path has been identified we can start to look at the detail of its interaction with the signal.

CHAPTER III

LEAVING THE TRANSMITTER

Radio frequency power from a transmitter must have an antenna to launch it on its journey into space. Here as everywhere there are losses, potential and real. Therefore, let's say that the transmitter feeds a transmission line, well matched at both ends, with an attenuation, hopefully small, that we know. We should not forget possible mismatch losses, but those we will put outside our present scope. The antenna will have a gain that we will need to understand, but it also has an interesting matching problem that we must somehow allow for, as its job is to couple the signal path to free space, which has its own characteristic impedance.

Antennas produce electromagnetic radiation fields, but within a few wavelengths of the radiating system the electric and magnetic fields must be separately considered. However, they are only very significant in the *near* field, where for example, they control personal EM exposure levels close in to the antenna. The radiation field provides the answers, in the *far* field, for greater distances,[1] and this is the field which is important for communication. All the fields decrease with distance, and so it is necessary to set some arbitrary location at which we hand over from local antenna system operations to a space propagation regime. So that we ensure that we are just dealing with far fields, we fix this at 1000 meters, sufficiently distant from an HF antenna to ensure that near fields are negligible. We also have to make a few assumptions, which we can modify later to model our real world. First, the antenna is isotropic; not quite the same as the proverbial vertical that radiates equally badly in all directions, as in that case we really mean only that it radiates equally to all points of the compass. The isotropic antenna (which is a convenient imaginary device that no-one could build) radiates equally well up, down, sideways or in any other direction. Next we place this imaginary antenna not in free space, but on the ground in our station's location, and provide our station with a perfectly reflecting ground surface. It's always easier to back out of such assumptions than to back into them at a later stage in the proceedings! All of this leads us to a very simple result for the field at 1 kilometer distance. [2]

$$E_0 = 5470 \sqrt{N}\ \mu V/m \quad \text{or} \quad E_0 = 74.8 + 10 \log_{10} N\ dB\mu V/m$$

where the field E_0, at 1 Km, is expressed in microvolts per meter or in dB relative to 1 microvolt per meter and N is the power in watts actually being radiated. When we substitute a real antenna, N will be the power in watts being radiated in the appropriate direction. The constants in the equations above, 5470 or 74.8, (74.8 being 20 $\log_{10}(5470)$) are determined by that small problem mentioned above, that the antenna must match into the 377 ohm characteristic impedance of space. Field strengths are usually specified in voltage per unit length. What this means is that if we could measure the voltage between any two points on a short piece of wire (short compared with a wavelength) located in the field, then if they were 1 meter apart in a field of 1 volt per meter, we would measure 1 volt of RF potential difference, and we would measure 0.1 volt if they were 10 cm apart. But in general, received radio fields are so small that we measure in microvolts per meter, or at most in millivolts per meter. If it's expressed in dB, the voltage is relative to a reference level of 1 μV/m, so 100 μV/m is the same as 40 dBμV/m.

Now that we know what sort of a field our imaginary isotropic antenna will lay down, at least at the start of its journey, we need to ask what a real antenna will do over real earth, and for this purpose it is convenient to enlist the assistance of an antenna computation program such as ELNEC or its bigger brothers EZNEC and NEC. These programs will provide an antenna gain in dBi, that is the gain relative to an isotropic antenna, exactly what we need. Although it's easy enough to decide that we want the best forward gain that our antenna will produce, before we select that figure let's look at the gain vs. angle of elevation. It's soon clear that the gain can vary drastically as elevation angle changes, and fortunately we were able to establish that angle when we calculated all the data associated with the great circle path. The angle varies with the path length, with the height of the reflecting layer and also according to whether our signals are received from the

E layer, the F layer, or even from a combination of the two. And this assumes that both stations are at ground level too! We will continue to make that particular assumption, although we'll want to ask our antenna program about gain variations that might occur with antennas at any reasonable tower or mast height.

We need therefore to have available to us, either in the form of a set of tables to look up, or as an accessible algorithmic calculation, the gain in dBi for the antennas we are interested in, at the height at which they may be installed, and at the whole range of possible ray angles for outgoing signals! (Clearly we will also have to use similar results later in the process at the receiving station.) It would also be great if we could include the effects of local terrain, both in ground conductivity and in configuration, but this might be getting a little too ambitious! Let's content ourselves with noticing that at most station locations, surrounding buildings, trees and other physical obstacles limit our antenna's view of the horizon, and so we should not become too excited about the possible benefits of extremely low ray angles, attractive though they may be.

Now that our signal, modified appropriately for the feeder system and for the characteristics of the antenna in its surroundings, has arrived at the 1 Km hand-over location, we can work on the next stage. But first we will look at ground wave propagation. Once we have a picture of that particular mode and its practical HF applications, we'll move on to examine the sky wave's first interactions with the ionosphere on its way to some final destination. However, before getting too deeply into the process, we will want to establish that the radio route is at all likely to exist. Are we, at least, below the maximum usable frequency?

Notes and References

1 Jordan p 314, 321
2 Rawer p 288

CHAPTER IV

GROUND AND OTHER WAVES

The earliest radio transmissions were over very short distances, usually less than 100 miles or so, and entirely by ground wave, the ionosphere playing no part at all. It came as a great surprise to most of the scientific community to find in the first decade of the 20th century that ionospheric propagation existed at all.

Ground waves comprise a surface wave and also a space wave, made up in its turn of a direct wave and a ground-reflected wave, distinctions which prove to be more important at VHF and UHF than at HF and lower frequencies. The surface wave is guided along the surface of the earth, much as any electromagnetic wave is guided by a transmission line. Earth is a distinctly imperfect lossy conductor and the surface extracts energy as the wave progresses, path attenuation depending directly on the earth constants of resistivity and permittivity. The space wave includes energy components resulting from diffraction round the earth, seen here as a large conducting obstacle, and from tropospheric refraction, due to changing pressure and temperature and therefore changing density of the air as the signal traverses higher altitudes. Under standard atmospheric conditions the line-of-sight distance for the space wave is increased about 15% by refraction. [1] Tropospheric reflection can also occur at any abrupt changes in refractive index, although again, this and associated ducting phenomena only become important at VHF. But note that not one of these mechanisms relies on atmospheric ionization or conduction.

The mathematical basis of ground wave propagation was derived very early in radio history, [2] although the equations and their solutions are so complex that semi-empirical relationships are used in most practical work. [3, 4] But important results follow from the basic theory. In the Sommerfeld zone, over land and away from the immediate area of the antenna, HF signals fall off at about 40 dB/decade of distance. In vertical polarization, attenuation is much less than in the horizontal mode, in which signals become quite negligible at very short distances. Medium frequency (MF) broadcast stations therefore rely on vertical antennas for their primary ground wave service areas. As a comparison, losses using horizontally polarized antennas can increase by about 50 dB over vertical polarization. For mixed polarization, losses are due both to propagation and cross-polarization losses at the antenna. Even simple horizontal dipoles in the real world radiate some vertical component, and computer modelling of a few normally imperfect locations with guy wires, masts and other conductors near the antenna, shows that 25 dB cross-polarization loss would often be a fair estimate.

Figure 4.1 Groundwave Propagation Curves Radiated Power 1 kW, over Good Ground.

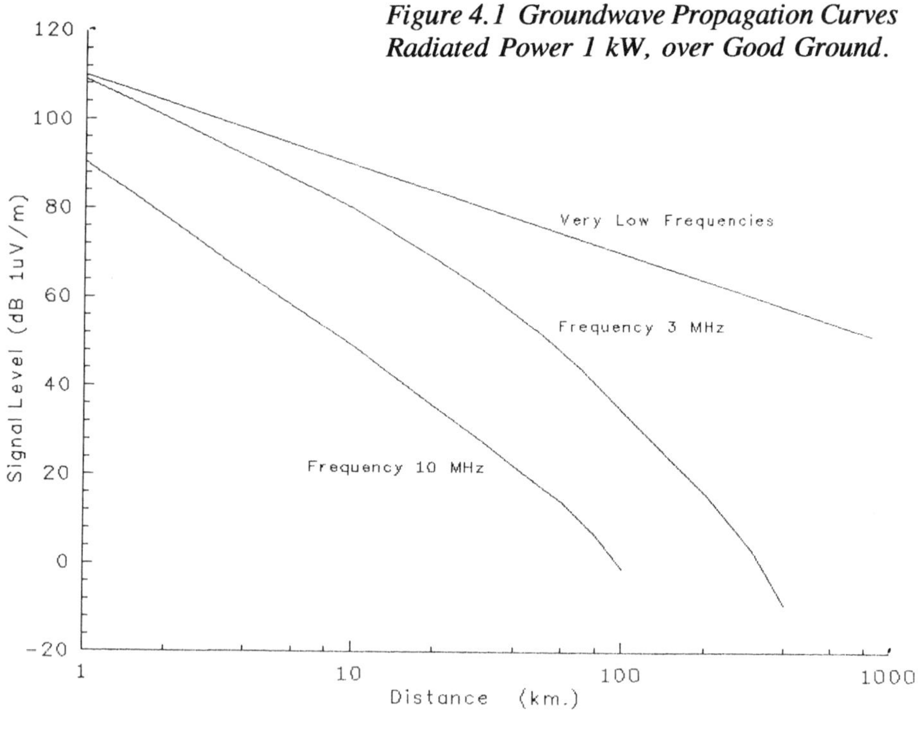

GROUND AND OTHER WAVES

The value of HF ground wave is in very local communications, except out at sea where salt water's high conductivity permits longer ranges. Data on ground wave propagation is presented in CCIR Recommendation 368.[4] A pictorial view may also be useful. Vertically polarized wavefronts, dragging their feet in the resistive earth, tilt further and further forward as they go; more so at high frequencies, but important for receiving antenna design in lower HF bands. If increasing tilt angle compares with the angle due to earth's curvature, then tilt effect tends to cause the wave to follow the earth's surface, but at higher frequencies, greater tilt can account for rapid disappearance of the ground wave as it tilts right into the earth!

Other modes are useful in the highest HF bands. Tropospheric forward scatter or atmospheric turbulence diffusion mode has very high point-to-point attenuation, 40 - 100 dB over free-space transmission. However signal loss is proportional approximately to fourth power of distance, rather than diffraction's exponential decay, so signals are much stronger beyond the horizon than diffraction theory predicts.[5] With a vertical or horizontal half wave dipole antenna and a 1 Kw transmitter at about 300 Km in flat terrain, signals might be -33 dBμV/m for 99% of the time, -12 dBμV/m for 50%, and 5 dBμV/m for 1% of the time. Rapid fading usually requires diversity reception. Controlling noise is often from cosmic sources.[6] Ionospheric scatter, optimally at 30 - 40 MHz, and auroral back-scatter support some HF communication, but need high power for reliability. Meteor trail scatter is useful for periods up to a few seconds at frequencies above 20 MHz.[7]

Notes and References

1 David & Voge p122-129

2 Jordan. Classic ground wave papers were by Sommerfeld and by Norton.

3 Terman

4 *Ground Wave Propagation Curves for Frequencies between 10 KHz and 30 MHz*, CCIR - Recommendation 368-5 Los Angeles 1986, also Braun

5 David & Voge, p143-151, JTAC 1964 p191

6 Saveskie, p144-7

7 Radio Communication Handbook, ARRL Handbook, Picquenard, p144-147

CHAPTER V

THE EXISTENCE OF A SKY-WAVE PATH

A signal leaving our earth-bound transmitting antenna is either headed for outer space, or as in the case we have just looked at, follows the surface of the earth. Ground waves, we saw, have a quite limited range, and for HF signals to reach any remote spot on earth, higher angle rays must be reflected back down to the surface. Before 1901, when Marconi's signals first crossed the Atlantic Ocean, it was generally expected that radio communication would be limited to a ground wave mode. His long span provoked considerable thought. Soon after that 1901 event, ionized reflecting layers were independently proposed by Oliver Heaviside and by A.E. Kennelly, although even in the 1880s[1] before the radio era began, several geophysicists felt that such layers might, or maybe must, exist.

For short-wave communications in the high frequency bands from 1 - 30 MHz, the important layers are designated D E and F, the F layer sometimes splitting into two separate layers F_1 and F_2. The D layer, at a height of 60 - 90 Km, is not considered permanent as ionization only occurs at a given location while the sun is actually shining there. Moreover, its electron density (D layer ions are electrons) varies considerably with time of day, season and latitude. The E layer, which exists at a fairly constant height around 110 Km, is also a daytime participant, although there is a very small residual night time ionization (just a few percent of the daytime maximum) which has been attributed to small meteors. It can support mid-length communication, but it also can shut off access to the F layers and prevent longer distance work. An even more ephemeral layer exists at the E layer altitudes; this is known as E_S or "sporadic E" and is generally only a nuisance in HF operation when it too prevents access to the F layers, but it can provide most interesting opportunities for the VHF operator. The uppermost or

F layer, the most useful reflector, varies greatly in height and ionization density. At local noon it is near maximum ionization and minimum height, some 250 Km. In summer, with strongest insolation, it rises and splits. F_1 rises just a little, while the F_2 component can rise as high as 400 - 500 Km, and there are other variations with latitude and time, not all of which have been very convincingly explained.

Now we must see if these layers will support communication on and at a desired path, time and frequency and at a given solar flux or sunspot number; predictions mainly based on ionospheric atlases. [Sunspot number and 2800 MHz solar flux correspond strongly despite slim physical similarity, but are so well described elsewhere[2, 6] that we just note that WWV broadcasts the flux hourly!] For vertical transmission, returning to a receiver on the same site, the critical frequency f0 is a maximum which the layers will reflect. And to get from critical frequency to MUF for a real point-to-point path, we apply the "secant law" which says

$$f_{MUF} = f_V . \sec\varphi_0 \text{ MHz}$$

where the angle φ is the ray's oblique angle of incidence at the reflecting layer. Typically we see critical frequencies for the E layer f0E, for the F layer, f0F, and especially for the upper or F_2 layer $f0F_2$. The E layer values are fairly regular, and can be quite well approximated by a reasonably simple equation. The F_2 layer however is quite another proposition with a very complex distribution, needing either an atlas or a small army of equations. A complete determination of MUF will also account for critical frequency depression due to ionospheric storms.[2] World-wide critical frequency maps have been compiled from many years of ionospheric soundings and for decades they supported laborious hand calculations assisted by graphs and tables.[3, 4] Dedicated slide rules have also been used,[5] but today computer programs carry this load.[6] The maps can be digitized, stored and used directly, but more often the mapped data has been reduced to a series of empirical algorithms. The first such program to be used extensively by radio amateurs was MINIMUF;[7] another very well regarded MUF predictor came from R. Fricker of BBC External Services.[8] These and similar routines have provided the MUF basis of other programs often strongly biased toward producing graphic route maps and signal strength displays.[9]

The critical frequency or zero-MUF, that is, the MUF for signals reaching up vertically to the ionosphere and returning to the original site at zero ground distance, provides another important insight. We have seen that MUF increases with distance, and if the frequency in use is higher than the zero-MUF, then our signals must meet the ionosphere at a low enough angle that reflection can take place, and therefore cover a greater ground range before they can be received on the ground. This greater range is known as the skip distance, and the skip zone, in which no reception is possible, extends from the outer edge of the ground-wave zone to the first point at which sky-wave signals can be received. It therefore ceases to exist as transmitted frequency is reduced to approach zero-MUF value.

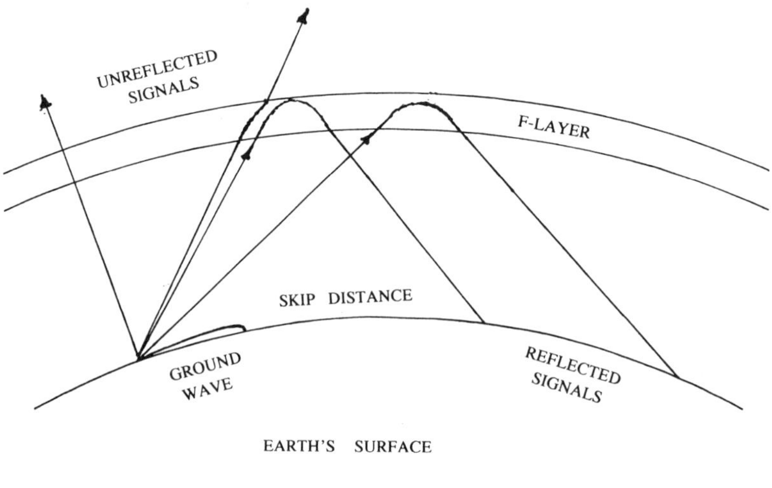

Figure 5.1 Ground Wave, Sky Wave and Skip Zone.

In commercial practice, MUF also provides the FOT (a French acronym for the optimum working frequency), usually calculated quite simply as 0.85 of MUF. The MUF defines, statistically, a sharp cut-off beyond which there will be no propagation, no matter what equipment or power levels may be in use. Although this is an overstatement, it certainly is in strong contrast to the other end of the HF range where the lowest usable frequency, or LUF has a relatively gradual onset and is very strongly dependent on equipment and systems. The LUF defines the lowest frequency below which the apparatus in use cannot communicate, but is not a cut-off in the same sense, rather a region where signal strength gradually falls until it no longer competes effectively with ambient noise. There are many cures. Higher transmitter power, higher transmit antenna gain, better local noise rejection at the receiver, a more tolerant modulation scheme; any of these can extend LUF to a lower frequency. Re-stating this in terms of transmitter output; increased power does little to raise MUF, but can do much to lower the LUF. [10]

Notes and References

1 100 Years of Radio, IEE Conference Publication Nr 411, IEE London 1995

2 Davies, NBS 80 Critical frequency depression, p266

3 Saveskie

4 Ostrow, NBS 90

5 *Slide rule for the calculation of absorption loss and of MUF in the lower ionosphere*, Barclay, L.W., Proc. IEE, 110, 9, 1523, 1963; also see Jones

6 Goodman

7 Rose

8 Fricker

9 HF propagation programs include SNAPmax, CAPMAN, IONSOUND, MINIPROP, and others, some only calculating maximum usable frequency. SNAPmax emphasizes the effects of noise and of modulation schemes for amateur radio and short wave listeners, in line with the scheme of this book, CAPMAN front-ends very comprehensive but not so friendly IonCAP, and MINIPROP is a general purpose program with good graphic route maps.

10 Jones

CHAPTER VI

THE SIGNAL ON ITS ATMOSPHERIC JOURNEY

Certainly the most interesting, and hopefully the largest losses the signal will encounter are those due to the radio path itself. These include distance loss due to spatial spreading of the ray, absorption in the ionosphere, and reflection losses.

Spreading Loss

After leaving the vicinity of the transmitting antenna, the radiated signals move out into space, and as they go farther and farther from their source so their power density is spread thinner and thinner and the resultant field strength at any receiving antenna they encounter will be less and less. The power flux would in fact fall off as the square of the distance from the transmitter if the environment were free space. But it's not, and if sky-wave communication is to be between two earth-bound stations, then the rays must undergo at least one reflection and often many more. This introduces a major complication; the reflecting surface of the underside of the ionosphere is not flat, but is the inside of a sphere, and the surface of the earth as a reflector is the outside of a slightly smaller sphere.

It's as well at this point to re-emphasize that the "layers" of the ionosphere are not layers in a physical sense, but rather areas of inflection in electron or ion density. Nor are signal rays reflected from them as in a mirror, but more subtly bent and guided back to earth. Nonetheless, for the purposes of our study, the concept of reflecting layers is not a bad one and we will continue to use it, remembering that if we want to study the actual mechanisms of the ionosphere, we very rapidly need a better picture. We should also note that the "reflecting underside" can be far from smooth, and is often not too truly spherical, and that all of these factors can lead to problems, and a few of them to opportunities, in propagation.

Concave spherical surfaces, the layer undersides, focus their reflections, while a convex surface, the earth itself, defocuses, and the combined effect after several reflections is complex. Kenneth Davies' classic study describes the important effects of focusing in many circumstances, [1] but goes on to note [2] that in the methods used by CRPL (Central Radio Propagation Laboratory) it was considered too complicated to take focusing effects into account in the calculation of spreading loss. A nomogram was suggested, which could be represented by

$$L_{dist} = 32.45 + 20 \log_{10}(f * Sdist) \text{ dB}$$

where f is frequency in MHz, and Sdist is slant distance in Km. Clearly the signals must travel farther than the point-to-point distance along the ground, and we describe distance traveled, including all the ups and downs as slant distance.

Others have also derived equations for spreading loss, and Rawer presents one of the most attractive descriptions, used by the French DPI (*Division de Prévisions Ionosphériques*). [3] Although Rawer's curves were originally presented in terms of ground distance, they can as in Figure 6.1 easily be translated to slant distance, which proves more convenient for calculation purposes.

Focusing Gain

Rawer indicates the possibility of focusing gain, one of the few areas where the signal can actually be enhanced in its passage. This gain can be real for paths very close to antipodean, although other mechanisms become important at such long distances, including round-the-world (RTW) routes. [4] A moment's thought shows that all possible great circle routes from transmitter to receiver can then potentially contribute at once to the final signal, whether from the front, back or sides of the transmitting antenna. While this gain can be quite large, it does depend on an unusually well-formed stable ionosphere and is not an effect to be relied on.

Ionospheric Absorption Loss

During each of its journeys from the surface of the earth to the reflecting E or F layers and back, the signal meets another part of the ionosphere, the D layer, at an altitude of some 60 - 90 Km. It plays no useful part in HF propagation, being altogether too thin and tenuous. But, like that famous river which is too thick to drink and too thin to plow, the D layer cannot be ignored as it absorbs energy from radio waves passing on through. And similarly, if the E layer is not ionized sufficiently to be a good reflector at our working frequency, it too will absorb wave energy. And as if this were not enough, the actual reflecting layer, E or F, (F in fact consisting of F_1 and F_2 zones), will also absorb a little, although usually this loss is only significant near the maximum usable frequency.

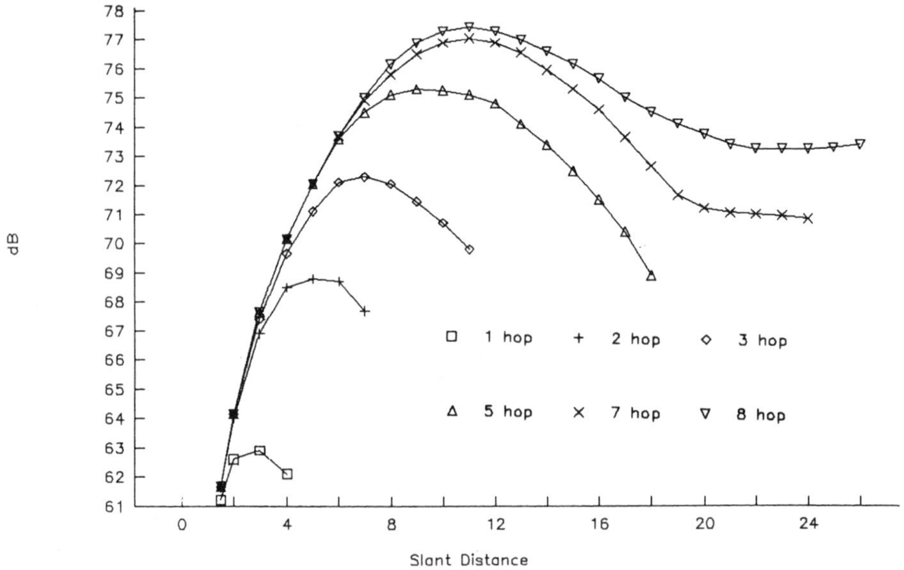

Figure 6.1 Path loss vs. slant distance x 1000 Km. (adapted from Rawer)

Again, there have been quite a number of different methods developed for the calculation of these losses.[5] They basically depend on the angle which the ray makes with the absorbing layers, the number of times it must pass through the layers, the ionization density in the layers, the frequency of the wave being absorbed, and gyrofrequency. Gyrofrequency describes an electronic dance routine that mainly affects the lower HF region and is explained in a little detail below. The ionization density is a function of the effective strength of the sun's radiation (it is mainly the sun's ultra-violet radiation and X-ray emissions which cause the ionization in the first place) falling on the layers at each point of passing, and this is calculated from the sun's zenith angle, which depends on local time and season at the location, and is given by

$$\chi = \arccos(\sin\Phi.\sin\delta + \cos\Phi.\cos\delta.\cos h)$$

where Φ is geographic latitude, δ is solar declination and h is the local hour angle of the sun measured to the West from noon. (Actually noon is fully specified as mean noon corrected for the equation of time and the standard time used at the location, which assumes higher precision than might be usual for operational radio, and certainly for most amateur interests!) Many years ago CRPL published curves for determining the zenith angle for each month of the year[6] and these have been widely reproduced. The CRPL equation for absorption loss is

$$L = (615.5 \text{ n } \sec\phi\ [1 + .0037R]\ [\cos .881\chi]^{1.3}\ /\ (f + f_H)^{1.98} \quad \text{dB}$$

although Davies notes that it is inaccurate below 5 MHz. The multiplier of 0.881 applied to χ relates the ground level solar zenith to the earlier sunrise and later sunset at D layer altitude. The number of hops is represented by n, the angle of incidence by ϕ, the sunspot number by R, the solar zenith angle by χ, frequency in MHz by f and the gyrofrequency, also in MHz, by f_H. This term is readily obtained from world gyrofrequency maps,[7] but here let's digress a little so as to see what it represents.

The ions of the D layer, which are in fact electrons produced by solar radiation in the far ultra-violet and soft X-rays, are subject among their other misfortunes to the earth's magnetic field, and the joint effect of impinging radio waves and that field is to make them vibrate at a characteristic frequency. This can be shown to be

$$f_H = eB / 2\pi m \ \text{Hz}$$

where e is the charge on the ion, B is the earth's magnetic field at the altitude of the layer and m is the ion's mass. For typical values encountered in the ionosphere the resulting gyromagnetic frequency is around 1.4 MHz. Motion of the ions in the layer, induced by the passage of a radio wave, is equivalent to the passage of a radio frequency electric current, and as the direction of motion is modified by the ambient magnetic field, the apparent dielectric constant in the region decreases, modifying the strength of the current. At low frequencies, close to the gyrofrequency, considerable additional attenuation can occur because of these effects, [8] and several other factors in the propagation equations are thereby slightly modified.

Another absorption loss formula given by Rawer [9] was based on a series of absorption tests made by the Radio Research Station at Slough in England under Sir Edward Appleton, and states that

$$L = p.B.\ \sec\alpha_D\ .(f \pm f_L)^{-2} \ \text{dB}$$

where p is the number of passes through the absorbing layers, $\sec\alpha_D$ is a "factor of obliquity" for a D layer at 80 Km altitude, f is the operating frequency MHz and f_L is the longitudinal component of the gyrofrequency, again in MHz. The magnetic component B is given by

$$B = 430\ (1 + .0035R)\ \cos^{0.75} \chi$$

where R is the sunspot number and χ is the solar zenith angle. The constant 430 represents an equivalent vertical ray absorption at 1 MHz, the coefficient of the sunspot sensitivity is just slightly less than used by CRPL, and the cos χ term has a rather lower exponent.

More recently this approach has been updated and refined by George and Bradley, who rely on a world-wide map of vertical absorption, [10] and like most more recent methods, it is just a little more complex. So, if you are trying to take a quick overview, you will probably want to skip over much of the next page.

Vertical absorption is that suffered by a wave propagated vertically from a transmitter site, most usefully in the form of a pulse which is reflected back from the upper layers (E or F) as in an ionosounder, [11] and received in the same location on its return. This variable is mapped worldwide as the equivalent absorption of a 1 MHz wave when the sunspot number is zero and the sun is located directly overhead. In this method the loss is stated as [12]

$$L_0 = A_0.(1 + .0067R)\ \sec\phi_{100}.\Psi(x).F(\chi).(f + f_L)^{-2}\ \ \text{dB}$$

where A_0 is obtained from the absorption map just described (and is a counterpart to Rawer's constant 430 dB), R is the sunspot number, ϕ_{100} is the angle between the ray and the vertical at 100 Km elevation, and the absorption factor $\Psi(x)$ is given as a function of x where $x = f.\cos\phi_{100} / f_0E$ and f_0E is the E layer critical frequency. The term f_L is the longitudinal component of the gyrofrequency and is equal to $f_H.\sin I$, where I is the dip angle of the earth's magnetic field.

> We often tend to think of the magnetic lines of force around the earth as going past us in a horizontal direction on their way to the pole, but a rather better model to have in mind is that of a big bar magnet buried along an axis near to the center of the earth. The lines of force at the equator will be approximately horizontal, but anywhere else they are apt to head downwards toward the nearest pole. The downwards angle is known as the dip angle and can be demonstrated with a magnetized needle which is free to swing up and down as well as around like a compass. Such a needle, or one which is only free to swing vertically (known as a dip needle) will show the dip angle.

The term $F(\chi)$ is equal to $\cos^p(0.881\chi)$. Here χ is the solar zenith angle, and $F(\chi)$ has a minimum value of 0.1, which allows for residual night-time absorption. The exponent p (which is Rawer's 0.75) is given by another map as a function of a modified dip angle. This composite value is derived from the geographic latitude and the magnetic dip angle [13] and is given by arctan(I / √cosλ) where I is the dip angle and λ is the latitude. You may also note that sensitivity to the sunspot number is somewhat changed. Typical values of these coefficients are given by Davies [14] as A_0 = 400 dB, p = 1.5, Ψ(x) = 1 and χ = 55°.

As we have seen, the "reflection" process at the upper layers should be thought of as a refraction or at least as a deflection, and one consequence of this analysis is an explanation of the observed result in certain ion sounder traces, that the wave splits into two components known as the ordinary and the extra-ordinary rays. [15] As the extra-ordinary ray has been shown to be subject to significantly greater losses at the reflecting layer in comparison with the ordinary ray, we can usually for practical purposes consider only the ordinary ray. [16] And since the energy content entering the two modes is equal, then we can further assume as a simplification, a reflection loss from this cause of approximately 3 dB, and our loss calculations should take this into account.

Auroral Losses

Those HF radio users living in the more northerly parts of the world, and those few who inhabit the Far South, will know the importance of auroral absorption for very many paths, mainly during ionospheric storms in which the Sun dumps showers of energetic particles onto the earth's magnetopause. Very fortunately for our life and health, most of these particles flow on past our planet into the night, forming the magnetosheath and its long extended tail. However, quite a few make their way through the protective belts above us and then spiral on down towards the earth, forming visible and radio auroras in Sub-Polar regions, where the dip angles are such that the magnetic lines of force come down almost vertically.

This concentration of particles occurs mainly as a ring or oval between the geomagnetic latitudes of 60° - 70° and varies both in position and in extent according to the ionospheric storm level, which can be related to the A or K index broadcast by WWV. The activity of the oval results in the increased absorption of radio signals as they pass through at auroral altitude, (close to the E layer), and this additionally depends on the particle density. The density is also related to the A or K index, and to local time at the point of intersection of path and oval as the oval is not of uniform density around the earth.

To calculate the effect of such crossings, we must determine the likely position of the oval, [17] next estimate its effective density, and then obtain from our path great circle calculations those locations where it will intersect with the oval at the auroral altitude, so as to see how many times it must run the gauntlet. [18] Combining these values we can obtain just a very rough estimate of the likely results in terms of signal attenuation, but in line with our general philosophy, an approximate result is probably somewhat more useful than no result at all.

Other Losses

There is also some scattering of energy in the reflection area, as the ionosphere is far from being a uniform entity with a smooth underside. By and large, losses from this mechanism are presently unpredictable, and contribute to some of the variability which makes sky wave transmission both challenging and fascinating. There is too the possibility of ionospheric tilts, where the "under surface" of the layer reflects wave energy in some non-standard direction. Opportunities are here for chordal hop modes, where the signal propagates without any intermediate power-losing earth reflections, or even without intermediate traverses of the D layer, and these can result in some of the spectacular but often very localized openings that again add to the interest of the HF section of the spectrum. [19] It is also argued that long path routes are often enhanced by a similar mechanism. [20] However, neither of these effects is very readily predictable, although it is interesting and easy enough to estimate the result of eliminating the relevant intermediate losses.

Polarization mismatch is a very relevant element for line-of-sight VHF circuits, and was also of great importance when we looked at ground wave operations. For ionospheric HF circuits, shifts due to reflections make it a somewhat unpredictable will-of-the-wisp. Arguments for the inclusion of an attenuation of either 3 or 6 dB to allow for this have been made,[21, 22] but as such it amounts only to a relatively small one-time addition to total path loss.

Reflection losses at the earth's surface on multihop paths merit a look. Good conducting surfaces, such as sea water, reflect HF rays rather well, with losses typically less than 0.5 dB with any angle of incidence greater than about 10° and generally rising with increasing frequency and falling with increasing incidence angle. Waves which bounce off land masses on the journey fare much worse, and on poorly conducting ground they can expect to be attenuated by some 4 dB or more. The land loss pattern is quite contrary to that at sea; it rises steeply at lower frequencies and also rises as the incidence angle increases. To assess these losses we have to look at all earth reflection points on a map of the world and allocate losses accordingly, remembering that fresh water and ice as reflectors are just about as poor as dry land in comparison with the oceans. For longer hauls, it can be a worthwhile exercise to select a route with sea rather than earth landings, and this sometimes helps explain performance improvement on long-path contacts.

A rag-bag of less defined effects is known as Over-MUF loss, applicable when we operate at frequencies beyond the statistical MUF, and since much amateur work is in this area, it too is worth examining. An expression for this is[23]

$$L_{OM} = 130\ (f - muf)^2\ /\ muf^2 \ \ dB$$

which may reflect the amateur experience that signals on a given band, at least on 14 MHz and above, are better at the top of the sunspot cycle than in the depths.

Looking deeper into this, Jacobs and Cohen[24] reported data for approximately 160 observations of four different ionospheric routes at a range of differing sunspot levels. On first thoughts, we expect to find that with higher sunspot numbers or solar fluxes, D layer absorption would be higher; propagation improvements, if any, would be due to our ability to work at higher frequencies where absorption falls off rapidly, in fact, as the square of the frequency. However this is not what the data appear to show. Normalization of the record shows stronger signals at constant frequency with higher fluxes for all four 14 MHz routes, from Virginia to Japan and Australia and to Far Eastern Russia. Figure 6.2 below shows the improvement which was found, and a correlation coefficient of 0.62 shows real dependency, though clearly subject to other perturbing conditions. [25]. If we apply Bradley's over-MUF loss allowance, it seems to compensate for this flux/loss relation, which may imply that the stations observed were "pushing the envelope" of MUF in normal amateur manner, rather than clustering, as would commercial operators with circuit reliability in mind, around the FOT at about 0.85 MUF.

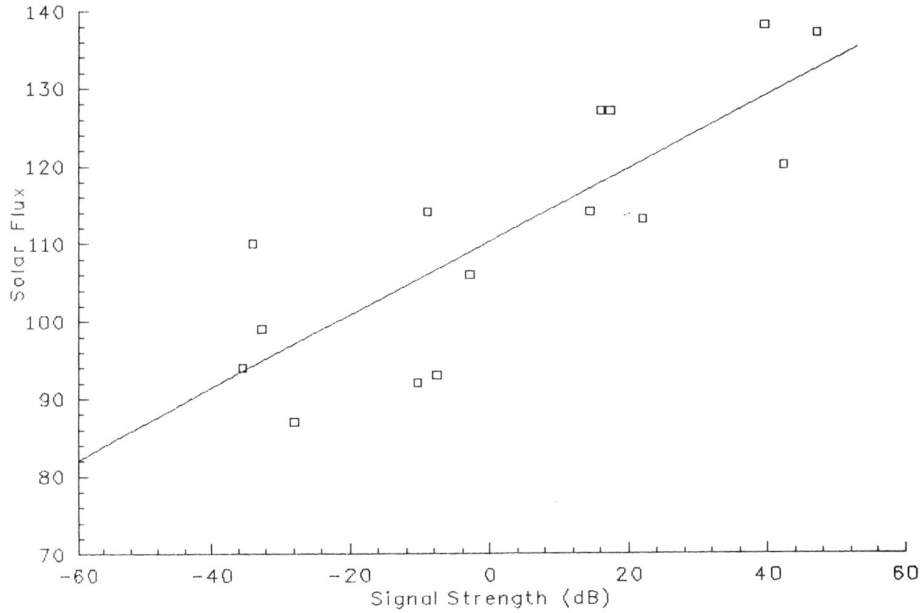

Figure 6.2 Solar flux vs. 14 MHz signal strength.

One further matter. Tropospheric weather (the kind we live in) has very little effect on HF signal strength, and can be ignored in our path calculations, although it certainly becomes a major factor in much of the the microwave region. It is however, as we are going to find, very important in the matter of HF noise level.

Normal Fading and Abnormal Fade-Outs

Sky-wave signals vary considerably in the long term, as solar UV radiation and flares change and occur, and as earth's daily and seasonal attitudes to the sun go through their normal cycles. But the very briefest exposure to HF listening will quickly persuade us that short-term phenomena are also at work, as signal strength fluctuates considerably from minute to minute, and even from second to second. As a rough guide to the extent of HF fading, let's say that average communication circuits provide a satisfactory link for about 90% of the time. For commercial operators it is often worth increasing power by about a factor of 20, or +13 dB to reach acceptable quality level. A 1000 times increase (+30 dB) would improve the 90% value to 99.9%, and yet a transmission level 30 dB down, only 0.1% of the original power, would still work out maybe 15% of the time! [26]

Most shorter term fading is due to addition and subtraction of wavefronts at the receiving antenna with slightly different path lengths and therefore different phase angles. [27] If only two waves arrive, with ray paths differing by half a wavelength, equal losses and equal amplitudes, cancellation will be complete and no signal will be heard. If paths differ by a whole wavelength the waves will add, increasing signal strength slightly. This simplistic view of a change maybe due to the motion of a reflecting layer, illustrates the common experience that fading can improve signals for a part of the time, but can cause a much greater loss of strength for a much larger fraction of the time. Higher frequencies mean faster fading. If a layer moves at 10 m/sec, this is 180°/sec phase change at 14 MHz, but is just 45°/sec at 3.5 MHz for a fading rate only a quarter as fast. Also as wavelength changes within sidebands, selective fading can occur, one modulation frequency fading independently of another, distorting demodulated audio signals. No energy is lost in *interference fading*, it is just redistributed among the receiving areas. Counter-measures are directive antennas favoring a single path, and diversity.

Polarization changes *en route* are important too, and can be non-reciprocal. For operations close to the maximum usable frequency, MUF failure, or changes in the MUF cut-off point, can also result in large variations in signal strength, typically for the short time-span "as the band is going out".

Dissipative losses from other geophysical mechanisms also cause irregular signal fluctuation, mainly during ionospheric storms and solar flares. *Sudden ionospheric disturbance* (*s.i.d.* or *Dellinger*) fade-outs occur maybe twice weekly at solar maximum and twice annually at minimum. Onset rapidly follows a flare or solar chromosphere eruption, sometimes with total signal loss over the HF spectrum, as soft X-rays and far U.V. emissions greatly increase D layer ionization. Even normal background noise may disappear, [28] though there is notable solar noise during s.i.d.s. Fade-out is usually earlier, more intense and more persistent on lower HF frequencies, often within one or two minutes, starting to return within just a few more minutes but recovering fully only after two or three hours. Only path segments in the sunlit hemisphere are affected, more so at lower latitudes. Less dramatic but more serious are ionospheric storms, which last longer, occur 20 to 30 times per year, and cause similar signal loss. [29] The resulting magnetic storms most severely affect the higher HF frequencies, up to normal MUF, with increased auroral losses on polar routes. [18] Ionization in the F layer is severely reduced, layer height increases, especially in polar regions, even topping 600 Km. Fading, often the fast *flutter* type, increases markedly. But transequatorial routes may be quite undisturbed, even enhanced. [30] The culprits this time are slow solar flare particle emissions which take about two days to reach earth's magnetosphere.

Yet another fade-out, *Polar Cap Absorption*, can be catastrophic in polar regions. *P.C.A.* is associated with flares but not directly related to either magnetic storms or aurora. High-energy proton bursts starting a few hours after flare onset may last from 24 hours to many days, causing very strong ionization down to lower D regions, far Northern and Southern path signal absorption increasing 100 dB or more. These events, rare at solar minimum but maybe monthly at high solar flux, are regularly reported by WWV. Ionospheric "weather" may cause other non-systematic changes. The D layer is by no means uniform; local irregularities associated with *traveling ionospheric disturbances (t.i.d.'s)* and *atmospheric acoustic gravity waves (a.g.w.'s)* can cause fading of D layer attenuated signals. Finally variable screening, wave diversion, or partial reflection, [31] can occur when rays split at sharp ionospheric discontinuities, typically sporadic E layer. If a received signal is reflected from that layer, total attenuation is quite possible, although signals passing through the layer may lose only a few dB.

Mechanism	Loss Range dB	Range	Period
Additive and Subtractive Interference Fading	0 to - 5 10 to 30	20 - 100 λ	1 - 1000 sec.
		2 - 20 λ	1 - 5 sec.
		0.5 - 2 λ	< 1 sec.
S.I.D.	to 60	Very large	Non-periodic
Ionospheric storm	Varies	Very large	Non-periodic
P.C.A.	to 100	Polar regions	Non-periodic
D layer variation	to 10?	50 - 500 Km?	300 - 3000 sec.
Focusing	to ±10	50 - 500 Km	50 - 5000 sec.
Diversion	Trans: 0 to 6	50 - 500 Km	10 - 500 sec.
	Refl: to 60		

Table 6.1 Signal fading distance ranges and time periods

Fading can be usefully described by its characteristic time periods and by the distance or range spacing of high and low strength signal reception zones. [32] Where the fade period is less than a second or so, (*flutter*), the cause is usually very local density and height differences in the ionospheric reflecting layers, often associated with polar paths and ionospheric storms. Slightly longer periods from 1 to 10 seconds are often due to larger motions of the layers. Fade periods of the order of a fraction of a minute to a few minutes can frequently be attributed to Faraday rotations of the polarization plane at reflecting ionospheric interfaces, while longer slower signal strength changes may be due to the onset or departure of favorable ionospheric tilts or focusing which can favor such as trans-equatorial routes. Very severe fading effects, familiar on MF broadcast signals, and typically much worse than for sky waves alone, occur at the skip zone borderlines where ground-wave and sky-waves have approximately the same strength.

The fading range describes the distance over which it takes effect. For example, Table 6.1 shows ray interference fading with a period of 1 - 5 secs. as estimated to be locally consistent over some 2 - 20 wavelengths; antennas for diversity reception may have to be separated by such distances. Diversity is clearly unlikely to improve reception under s.i.d., ionospheric storm and diversion fading.

Doppler Shifts and Multipath Problems

Where ionospheric motion is evident, there may also be Doppler shift of the received frequencies. The magnitude of such shifts typically does not exceed a few Hertz, unlikely to upset even narrow-band systems, although maximum values up to 20 Hz or so have been recorded. [33] Where the motion is irregular a single transmitted frequency may be received as a small range of varying frequencies. Often associated with Doppler shift is multipath propagation, with time-of-arrival differences of a few milliseconds, large enough to cause considerable inter-element interference with higher speed digital transmissions. Sometimes such interference can make even relatively slow 45 Baud RTTY quite unreadable despite excellent signal-to-noise ratios, an effect familiar to many 80 meter digital operators. Manual Morse code has a definite advantage here, as it is very easy to reduce sending speeds to compensate.

In general these effects most frequently occur on shorter paths, less than about 1000 Km, in the lower HF bands below 10 MHz, and typically near the lowest usable frequency (LUF), [34] and they are perhaps most familiar on polar routes under aurora conditions. Multipath is best avoided by operation as close as possible to the MUF. Criteria have been developed to predict the likelihood of multipath delays with frequency and path length, and troublesome interference is most likely when two paths produce similar signal strength, say within 10 dB. [35]

Digital ionospheric communications are limited even in the best possible case by received signal frequency instabilities caused by normal small-scale movements of the layers and by ionospheric roughness, so that the highest practicable speed for HF transmission is probably no more than about 10 Kbit/sec. [36]

Reciprocity

On the face of it, we expect radio paths to behave similarly in both directions, but this is not necessarily true, as experience occasionally seems to tell us. At least two different phenomena may be responsible. One which has been quite carefully studied is that due to polarization changes at reflections, maybe favored by some antenna polarizations.[37] In another mechanism, ionospheric "roughness" causes down-coming waves to be more nearly vertical than upgoing waves.[38] Especially on multiple-hop circuits, the effect being cumulative, the reflection sites for signals in one direction will be slightly displaced from those for the reverse direction, and the layer (or even terrestrial) characteristics at those sites could be sufficiently different to noticeably change observed propagation.

Notes and References

1 Davies, NBS 80, Ch 5.
2 Davies, NBS 80, 7.3.4.3
3 Rawer, p290, David & Voge, p273 also compares this with CRPL results.
4 Hortenbach and Rogler. [Very long distance, including RTW, propagation is treated in depth in *Long Distance Propagation of Radio Waves*, Gurevich, A.V., and Tsedilina, E.E., Springer Verlag, Berlin 1985]
5 Ostrow, Davies NBS 80, Rawer, Barghausen et al., George & Bradley, and CCIR 894, for example.
6 Ostrow, NBS 90, Figures 23-34.
7 Davies, NBS 80, Figure 7.7
8 David & Voge, p186
9 Rawer, p290,291
10 George & Bradley
11 Davies, NBS 80, Ch 3
12 George & Bradley and Davies, IEE p225
13 Davies, IEE p44
14 Davies, IEE p226
15 Davies, NBS 80, Ch 2

16 Rawer, p290 and *Field Strength Determination at High Frequencies*, Bradley, P.A., p395, see Jones
17 Goodman
18 Picquenard, p248
19 Brown, p68-77
20 Hortenbach & Rogler
21 Brown, p43
22 *Field Strength Determination at High Frequencies*, Bradley, P.A., p395; see Jones
23 Goodman, p254, Davies, IEE (note typographic error in formula.)
24 Jacobs & Cohen, 1982, p112-120
25 MacKeand, p9-16, with corrected graph in Fall 1997, p105
26 Picquenard, p154-155
27 Davies, NBS 80 p242-7, Goodman, p268-273, see also ref. 31
28 *Sudden Disturbances of the Ionosphere*, Dellinger, J.H., Proc. IRE, v25, p1253, Terr. Mag. & Atmos. Elec., v42, p49-53, Davies, NBS 80 p258-262 and Bennington, p126-130
29 RSGB Handbook, p12.13
30 Davies, NBS 80 p266
31 *Polarization and Fading near the LUF*, Hayden, E.C., p262-4; see Jones
32 Goodman, p269 & p317, see also ref 31
33 *The Calculation of Ionospheric Doppler Spread on HF Communication Channels*, Pickering, L.W., IEEE Trans. Comm., 23, 5, 1975 p526-537
Equatorial Spread F Effects on an HF Path: Doppler spread, spatial coherence and frequency coherence, Fitzgerald, T.J. & Carlos, R.C., Radio Science, v34, 1, p167-178
34 *Some Results of an Oblique-Incidence Pulse-Sounding HF Experiment between Andoya, Norway & College, Alaska*, Bartholomew, R.R.; see Jones
35 Goodman, p262
36 Goodman, p264
37 *Short-term and Averaged Characteristics of Nonreciprocal HF Ionospheric Paths*, Jull, G.W.; see Jones
38 *The Reflection Coefficient of an Ionosphere Containing Large Scale Irregularities*, Caldwell, J.D., Stevens, E.E., and Warren, E.S.; see Jones

CHAPTER VII

NOISE

When radio was yet young, "static" was the major competing noise. Although receiver selectivity was poor or lacking, few stations were transmitting, and the cacophony from other electrical equipment was much less widespread. On the other hand, the earliest radio techniques brought their own noises. Nothing can have made quite such a racket, on or off frequency, as the crashing signal of one of the major shore-based high power spark transmitting stations of the 1910 era!

But signals must always compete for the receiver's ear; sometimes the result is a walkover and sometimes the battle is lost. Noise is always the other team, a team with many members, some easily neutralized, some stronger or weaker depending on location, or time of day, or even time of year, and some just too much to handle. On-frequency station-to-station interference is usually today's greatest problem, due either to adjacent in-specification signals and insufficient channel spacing, or to spurious outputs from other stations, whether operating correctly or, too often, not. Many international conferences have attempted to minimize such interference by assigning frequency bands to particular services, although in the history of radio development, the pace of innovation, generally on new and higher frequencies, has often far outstripped any official efforts. [1]

The earth's atmosphere produces noise from thunderstorms, usually far-away but sometimes local, even too local, and on occasion gives us corona noise or precipitation static. The rest of the universe, including our sun, sends cosmic noise, although this is only rarely a real problem in the HF bands. Then there is man-made noise, from automobile ignition and from diverse local electrical operations and the power lines that feed them, sources that have grown by leaps and bounds. Unlike interfering radio stations, they have no communication intent, their radio frequency emissions being purely incidental to their operations. One example of this growth is today's need for a 2 MW LF broadcast transmitter to cover Poland, where 120 KW was adequate in the 1930s. [2] Finally, radio receivers also generate their own internal noise, but this is very rarely an important competitor at HF, and even then only at the higher end of the band.

Noise Competition and Tolerance

Some systems and services are fairly tolerant of competition, others are not. Commercial point-to-point communications need a predictable signal-to-noise margin and must provide service at times dictated by the user rather than the operator, but often have the possibility of selecting favorable sites and can frequently use fixed higher gain antennas. HF broadcasting must operate with unsophisticated receivers and receiving antennas with no possibility of receiver site selection and often there will be competing signals of somewhat equal interest. Amateur communicators can only rarely select their sites, (maybe most often in portable operation), operate at times largely selected by the operator, generally use low transmitter powers, and often seek to use paths that would be too poor for consideration in other services. On the other hand amateur stations can and do in many cases use fairly sophisticated receivers, antennas and modulation systems, and can use channels despite fairly frequent signal path interruption by the simple means of requesting repeats, manually or, today, automatically.

Noise level data is well established for many sources, but naturally tends to be directed to situations that can limit the economic operation of broadcasting, or point-to-point commercial work. The amateur service tolerates lower signal-to-noise ratios but is also interested in operation where low local noise levels may prevail. For example, it may be worth knowing how much benefit can be obtained from low-noise sites for portable work, or by operating a home station during periods of low traffic on an adjacent highway, or during fair weather if the station is unfortunately close to an EHV overhead line.

Station-to-station Interference

Co- and adjacent-channel broadcast interference has been quantified,[3] but is also a fact of life in point-to-point services. In the crowded amateur bands where no specific frequencies are assigned and the number of operational stations per KHz is the highest to be found anywhere, signals are all too likely to be received on our operating frequency from other stations whose legitimate bandwidth coincides in whole, or in part, with our own. It is only too common that a frequency is apparently clear, and yet we interfere with a contact, one half of which is quite inaudible to us. Technical remedies for this are limited; a directional antenna (in bearing, or maybe in elevation) may help. Operating practices ... look carefully first, move frequency, or frequency band, or choose another time if necessary ... are clearly important, but in the heat of a contest the only possible remedy may be courtesy. Sometimes a change of mode will help, maybe by simply decreasing required bandwidth, maybe by increasing inherent robustness. Where double sideband full carrier A.M. didn't make the grade, single sideband A.M. might; where Baudot RTTY does not, PSK31 may well; all presuming an approximately even playing field. But clearly one option is sending in twenty players to the other side's eleven! Antenna improvement is more acceptable to most than is a simple increase in transmitter power, but in an area of strong feelings and strong facts, we will duck these well discussed issues and consider other opportunities!

All radio stations radiate on non-required frequencies, though today's jurisdictions set limits. *"Emissions resulting from modulation must be confined to the band or segment available to the control operator"*, and *"the radiation of harmonics and other spurious emissions shall be suppressed to such a level that they cause no undue interference with any wireless telegraphy"*.[4] We can't do away with such outputs completely; one specification requires that spurious emissions shall not exceed 50 mW, and shall be 40 dB below the fundamental emission. Thousands of transmitters operate world-wide, maybe with a total intended radiation of some 10 watts/Hz of HF spectrum. Broadcast, marine, point-to-point, mobile, amateur, they cover virtually the entire spectrum.[2] There is therefore today a world-wide background noise from this cause, possibly 1 - 10 watts/KHz on any "clear frequency" we nominate, which did not exist a hundred years ago. The resultant competition at the receiver will depend greatly on propagation from major world centers, and must generally be lowest in the remotest parts of the world. (Although HF broadcasters and others do locate high powered transmitters in some very out-of-the-way places!) And of course, there may be spurious outputs that are large enough, from sources near enough, that they peak on specific frequencies which cause very severe problems locally. The amateur friend next door operating on 40 meters when we are trying to run DX on 15 or 20 meters, must be just about the extreme case of this situation.

The spurious continuum we cannot escape. Locally severe peaks are sometimes avoidable by selecting our operating frequency, or a time when the interfering source is not operating. Knowing the source identity, we might persuade the operator to shut down, or most desirably, correct spurious outputs where they are beyond legally permitted emission limits. This can also apply to line leakages from CATV for instance, although they normally only present a VHF problem.

Local Transmitter Interference

A strong local transmitter, or a multiple transmitter set-up as on a field-day site, can produce interfering signals far above and beyond spurious transmitter outputs. Modulation can transfer from an interfering signal to a signal on the receive frequency; an interfering signal can drive the receiver front-end into a non-linear overload condition; two interfering signals can mix in a receiver non-linearity, and of course receiver spurious responses can appear, such as "second channel" in a simple superhet, or direct reception on an insufficiently shielded intermediate frequency (IF) amplifier. Although some modes may be more tolerant than others, in all cases, sufficient shielding and sufficient front-end receiver filtering will eliminate the interference, provided that filter insertion losses are not so high as to eliminate the wanted signal! We can often rely sufficiently on the robustness of the receiver's front-end, although simple band-stop filters can often abate high induced voltages from neighboring MF AM broadcast stations, for example. Of all such problems, cross-modulation on naval vessels where multiple high powered radio and radar systems share a small platform, may well be the severest. [2, 5]

Most pernicious is the "rusty bolt" phenomenon, where strong local signals mix in unsuspected non-linearities to produce apparently unrelated interference. Also classically in ship-board installations, iron oxide junctions, for example in antenna staybolting, have produced very variable amounts of hard-to-track RF. On land, aluminum downspouting has also offered the same disservice. Clearly such problems, difficult and unpredictable as they are likely to be, cannot be laid at the door of the transmitter operator and must be resolved locally at the receiver.

Atmospheric Noise

The static crash of thunderstorms[6] is the inescapable natural background to radio, very much more insistent at low frequencies than high. In the lowest radio bands, 10 - 100 KHz, "sferics" have been used for thunderstorm tracking for many years. Lightning strokes are very large sparks, and as such produce large amounts of radio frequency energy. In a simplified view, two such sources can be usefully distinguished, cloud-to-ground discharges maybe with a path length of 1000 meters or more, and in-cloud discharges, usually with much shorter path lengths. The ionized path can be thought of as a radiating antenna, and so the cloud-to-ground discharges account for low frequency noise where wavelengths are maybe 1000 to 10,000 meters, while in-cloud strokes characteristic of tropical storms, are responsible for most atmospheric noise in the band 2 - 20 MHz. A typical in-cloud discharge has an overall duration of about 200 ms, and is composed of four stepped leaders each of which lasts about 1 ms. Many steps, each one lasting maybe 75 μsec, recur in each leader, an average step being some 67 m. long, a radiating length which places these sources squarely in the low HF bands.[7]

It is interesting to compare atmospheric noise power with that due to incidental transmitter emissions. We are told that in a bandwidth of 1 KHz, mean power in typical atmospheric discharges is 1.5 W at 1 MHz and 60 - 80 mW at 10 MHz. World-wide, about 100 lightning discharges occur each second, 60 or 70 being in-cloud. From the global gamut of thunderstorms we can therefore infer an equivalent RF generator at 1 MHz of about 20 watts/KHz, falling to 1 watt/KHz at 10 MHz, not too different in magnitude from previously noted spurious transmitter outputs.[8] (But at low frequencies, natural generators far outweigh man-made contributions.) Lightning noise is very variable locally due to its HF propagation, to the strength of the storm and to its location relative to the receiving site. The noise has a Gaussian statistical distribution, largely because it propagates from source to receiver over many different paths simultaneously. Interestingly, it also has a fractal nature, being self-similar over both long and short periods. Considering aural reception, we must set an integrating period for this series of crashes or impulses. Human ears appear to have time constants of about 10 ms in "charging" and about 500 ms in "discharging", so that any sound's full strength is estimated in about 200 ms,[7] and at lightning impulse frequency, audibility as interference can be compared almost volt for volt with signal level.

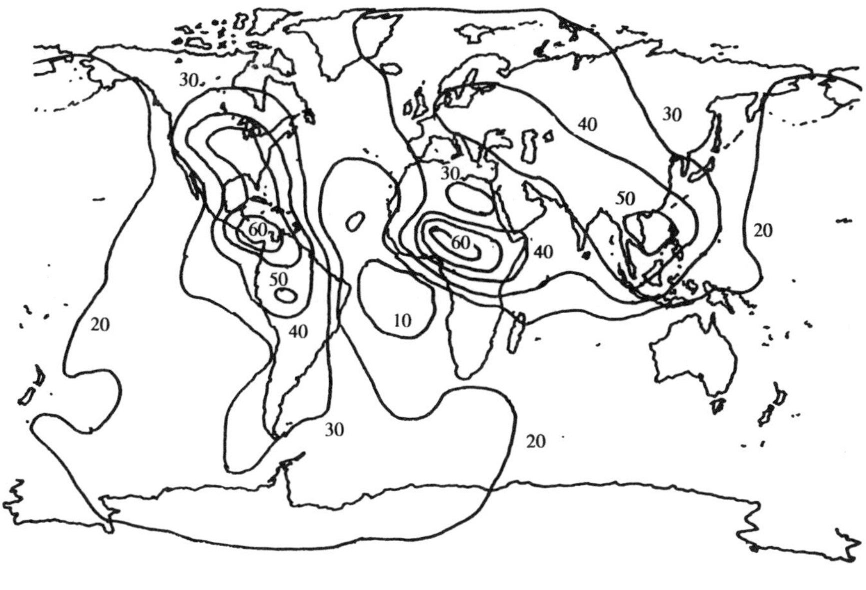

Figure 7.1 Noise Grades, June through August, 0800 - 1200 Local Time

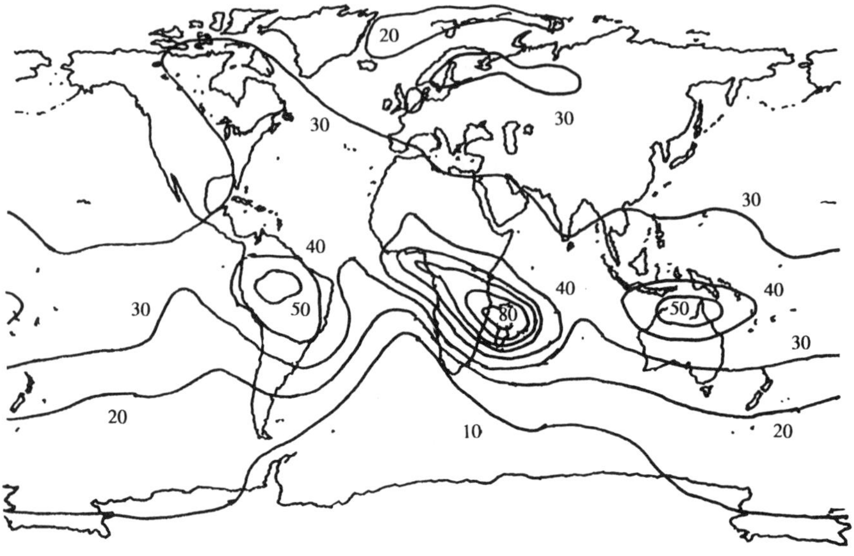

Figure 7.2 Noise grades, December through February, 0800 - 1200 Local Time

World-wide noise data was first compiled in England in the World War II period, at the Interservices Ionosphere Bureau, and was subsequently published by service units and the National Bureau of Standards. [9] By the 1960s sixteen stations with compatible recording systems in all continents and mid-Pacific contributed to the CCIR Report 322 data base. This comprehensive, widely reproduced set maps the world's noise grade contours at 1 MHz for six daily periods in the four seasons. (Northern winter is December - February, Spring is March - May, not the usual US convention of January - March etc.). Associated curves relate noise grades to other frequencies from 10 KHz to about 30 MHz. [10] Collected noise information alone was insufficient to produce useful noise maps, so the extensive existing meteorological storm records,[6] together with radio propagation computations, helped determine likely contours, noise data setting actual levels.[8] More recent publications have relied less on extrapolation from thunderstorm records.[11] The maps of Figures 7.1 and 7.2 use a slightly different format from Report 322, being by quarter rather than by season. Figure 7.1 (June - August), for instance, shows the noisy Northern hemisphere Summer with the relatively quiet Winter in the South to help visualize world-wide distribution. However, like the CCIR maps, they show noise data for blocks of local time, and not a world-wide noise snapshot at any one epoch. Figure 7.3 on the other hand does indicate a possible array of thunderstorm noise as a world-wide distribution for one time, one date, arbitrarily selected as 10:00 GMT in July, but it is worth emphasizing that all these maps show received noise levels rather than source strengths.

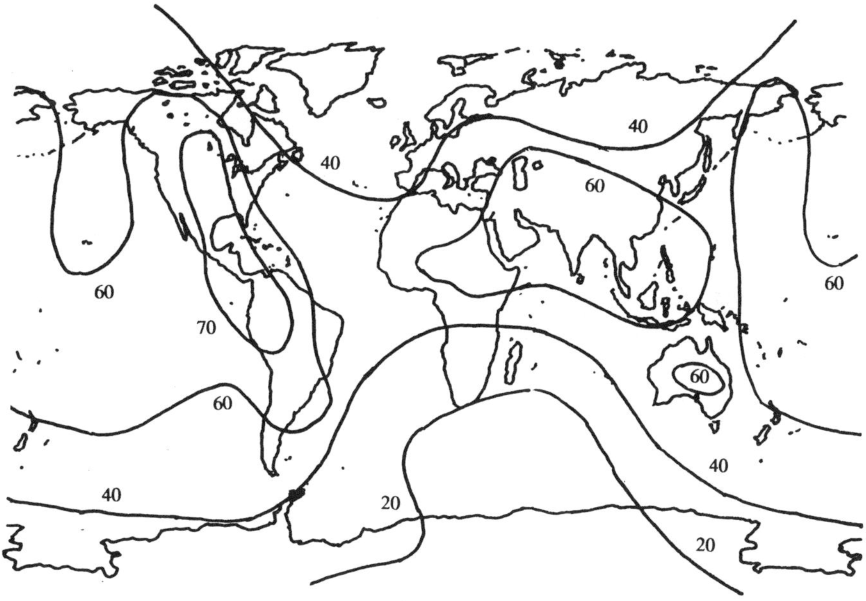

Figure 7.3 Noise Grades, a one-time snapshot, July, 10:00 GMT.

Noise grades are defined by mean noise power, generally agreed to be the most convenient and useful expression. Noise power available from a lossless antenna is expressed in dB above a reference power level kT_0b; k is Boltzman's constant (1.38×10^{-23} Joules/$^\circ$K), T_0 is reference temperature, here 288°K or 15°C, and b is bandwidth in Hz. At this reference temperature, $10 \log_{10}kT_0$ is 204 dB below 1 Joule, and so 0 $dBkT_0b$ = -204 dBW, i.e. 204 dB below 1 watt at a bandwidth b = 1 Hz. (1 Watt is defined as an energy flow of 1 Joule/sec). The noise factor F_a is then specified for a short vertical antenna on a perfectly conducting ground plane, and as such F_a relates to r.m.s. field strength induced along the antenna by

$$E_n = F_a - 95.5 + 20 \log_{10} f \;\; dBkT_0b$$

where E_n is r.m.s. noise field strength in dB above 1 μv/m for a 1 Hz bandwidth, F_a is the noise factor at the actual working frequency, and f is frequency in MHz. The field strength for any other bandwidth is determined by adding $10 \log_{10} b$, where b is bandwidth in Hz.

Atmospheric noise is far from constant in the shorter term. However, when a mean level is struck over a period of at least several minutes, average noise power is found to be fairly constant, usually within no more than ± 2 dB in any hour except at sunrise and sunset. In the low HF bands the dawn and sundown periods are characterised by rapid noise level changes, typically about 15 dB/hour; these changes are less pronounced at 10 MHz and can virtually disappear at 20 MHz. In the longer term, it may be surprising that no significant changes are seen with sunspot number. As worldwide noise sources neither increase nor diminish with the sunspot cycle, the total received noise at a given site remains fairly constant, although the pattern of noise propagation paths will certainly change. Another degree of compensation is also noted; as ionospheric propagation of atmospheric noise tends to fail, so also will the ionospheric screening of cosmic noise!

Typical HF antennas are directional, and HF atmospheric noise sources are fairly localized; we might therefore expect to be able to make allowance for this. However experiments in southern Australia with antennas beamed in various directions showed noise levels lower than standard short vertical noise antennas, while similar German tests showed higher levels. Tests in Singapore using a very directional rhombic antenna system showed noise reception on the rhombic up to 4 - 6 dB greater than on the vertical noise antenna. Vertical ray angles are probably an important factor in these differences. Allowances are difficult to assess and an unpredictable ±6 dB variability in this regard has been widely accepted, [8, 10] although in particular cases such calculations are possible. [12]

Relative polarization of signal and noise may also have some importance. E_n is measured as the vertical component of the noise field, but while mapped noise levels can be considered independent of polarization, to a first approximation, as long as antennas are high compared with wavelength, [10] data on this aspect is even scarcer than on directionality.

Report 322 gives mean noise grades, and their standard deviations, which are usually around 4 - 6 dB at HF. To calculate circuit performance, upper decile noise is often used rather than the mean, assuming that for some commercial purpose we would be satisfied with communications 90% free of noise intrusion. The curves below relate the mean to upper and lower deciles. The upper decile addend tapers off below 100 KHz, and decreases roughly linearly with logarithmic frequency from a high at 2 MHz to a low at 20 MHz. Summer values at 2 MHz are around 15 - 17 dB from 0400 to 2000 Local Time and 8 or 10 dB at other times. At 20 MHz they are generally around 6 dB at all hours. In Winter, the higher values generally pertain only from about 0400 - 0800 and 1600 - 2000. The deduction for the lower decile has a similar value, but is usually about 2 dB less than the upper decile addend.

While statistical circuit service calculations are a commercial necessity, in amateur communication, where by definition, link establishment cannot be a necessity, circuit quality must often be remarkably bad before established links are given up. The moment's need may be satisfied by 50% reception, or even a transmission loss close to the 90% lower decile would suffice, if only the remaining intelligible signal segments had the wanted data, often repeated in a friendly attempt to help. The unadjusted mean noise level is therefore arguably the best value for most purposes in this non-commercial arena.

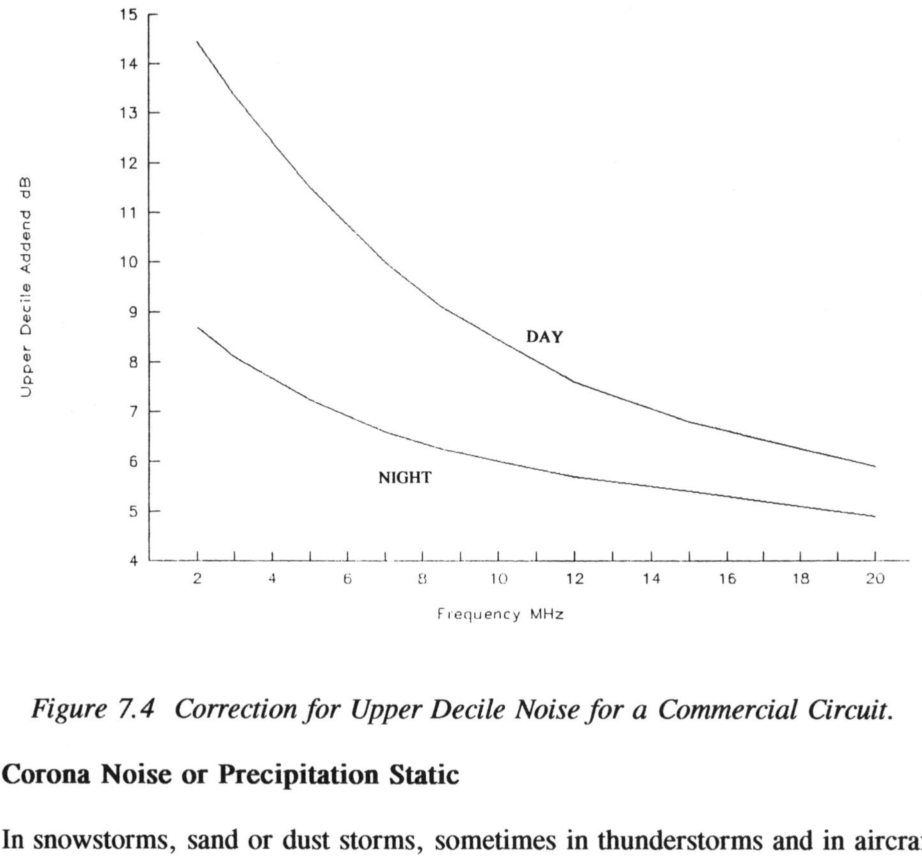

Figure 7.4 Correction for Upper Decile Noise for a Commercial Circuit.

Corona Noise or Precipitation Static

In snowstorms, sand or dust storms, sometimes in thunderstorms and in aircraft flying through rain or ice clouds, corona can appear on sharp points of antennas, discharging very high static voltages produced by particle impact or induction fields from charged clouds. Resultant RF noise, often a high pitched "scream" in one or two minute cycles, can completely blanket the HF bands, destroying any and all reception. [13] Avoidance of sharp points, special tips for whip antennas, and appropriate antenna grounding in ground stations, provide the usual remedies. EHV power line corona noise, also weather related, is considered separately.

NOISE

Cosmic Noise

Hot bodies radiate heat or even light, and the physics of "black-bodies" shows that they must emit radio waves as well. The sun and the galaxies of the universe are copious generators of electro-magnetic radiation, but our atmosphere has only two window regions which allow passage of that energy from outer space to earth itself. These are from a wavelength of 4×10^{-5} cm. to 7.2×10^{-5} cm., namely light and heat with some ultra-violet and some infra-red, and from about 0.25 cm. to around 20 - 30 m., so including the upper end of our high frequency radio region. In December 1931 Jansky discovered, to general surprise, that 15 m. radio waves were reaching the earth from space, as his antenna noise varied daily on a regular basis, both in direction and strength. The actual periodicity was 23 hours and 56 minutes, a sidereal day, the period corresponding to the earth's rotation relative to the stars, and on further investigation he concluded that the source was in the Milky Way, in the constellation Saggitarius.

The first directed scientific searches for solar noise, by Reber in 1940 and 1942, were inconclusive, but by 1944 the fact was established. In 1945 Appleton noted that radio amateurs in 1936, during solar cycle maximum, had reported a hissing sound in 10 - 40 MHz receivers, only in day-time and often as a precursor of the s.i.d. or "Dellinger" fade-outs associated with solar eruptions. If these were in fact of solar origin they had to be some 10,000 times stronger than Reber's black-body noise. In February 1942, British Army *GL* radar units also reported very strong noise on 4 - 6 m. wavelengths, feared to be enemy jamming, but this too proved to be of solar origin, and was presumed to be from an observed large sunspot. In 1946 Ryle and Vonberg measured such effects with one of the first radio interferometers on 175 MHz, and it was discovered that about a half of the major sunspots were substantial radio noise sources. Solar flares are also responsible for strong noise emissions, which occur together with auroral displays and geomagnetic storms. Clearly it would be highly edifying to find meaningful data embedded in all this radiation, but although there have been, and continue to be, many attempts to find such extra-terrestrial intelligence, so far all we have is noise and more noise! Which should not lead us to forget that a great deal of good scientific data has nevertheless been extracted from that noise. [14]

In ground-to-ground HF communication, cosmic noise is significant only at relatively high latitudes or at higher frequencies. An "*iris cut-off*" occurs, as the ionospheric layers reflect radio energy not only from earth stations back to earth, but also from space back into space from the layer topsides. The *iris* is an effective opening in the ionosphere for cosmic noise, small near the critical frequency but increasing at higher frequencies as $sec^{-1}(f/f_c)$. CCIR Report 322 extrapolates HF noise levels from VHF measurements, [15] showing that for simple vertical antennas, cosmic noise and man-made noise can have similar magnitudes in the higher HF range, and that cosmic noise can sometimes predominate if the critical frequency is below the frequency in use. The cosmic background level in practice is some 15 dB below the maximum encountered with directional antennas.

Man-Made Noise

In the most isolated places, far from cities or major industry, man-made noise can be negligibly low. Commercial HF receiving stations are therefore often located on remote rural sites. In urban areas however, levels may be high enough to make HF radio communication quite difficult. [2, 3] The sources and nature of man-made noise vary from the impulsive noise of automobile ignition systems to continuous hash from sparking commutators, or radio frequency leakage from such deliberate generators as HF heating or medical diathermy equipment. All of these can be radiated and received directly from the source, but some are more likely to be transmitted over power lines, with or without re-radiation to the receiver, the lines themselves also being responsible for some RF noise generation. [16] Generally, the two most identifiable sources are automobile ignition systems and power lines.

Automobile ignition noise was once a major problem but modern designs, (maybe mainly to accomodate the use of car radios!) have suppressed this source well enough that now it rarely presents real difficulty. Breakdown in an unsuppressed spark plug gap occurs in several phases, in one of which, a very high oscillatory current, peaking as high as 250 a. with a duration of some tens of nanoseconds, radiates noise to suit. [17] Series resistance, either as discrete resistors or as high resistance "wiring", reduces the peak to around 1 a., dramatically reducing noise

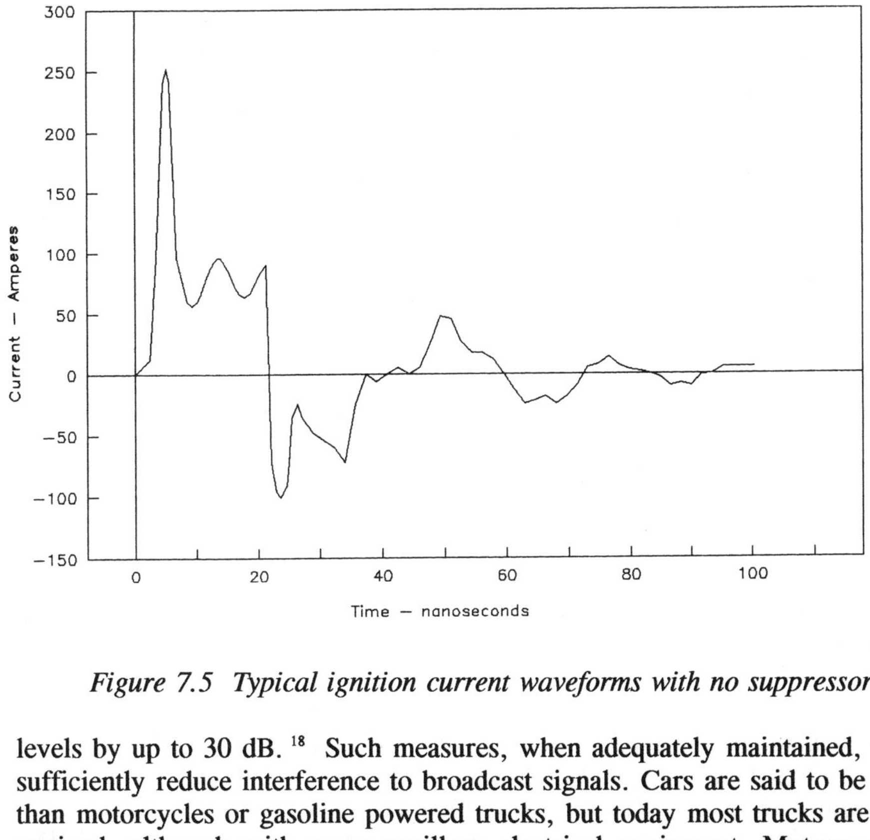

Figure 7.5 Typical ignition current waveforms with no suppressor.

levels by up to 30 dB. [18] Such measures, when adequately maintained, usually sufficiently reduce interference to broadcast signals. Cars are said to be quieter than motorcycles or gasoline powered trucks, but today most trucks are diesel-engined, although with more ancillary electrical equipment. Motorcycles are smaller but less well shielded, car bodywork reducing engine RF emissions up to 15 dB or so. Further ignition system shielding may also be needed to meet the needs of mobile HF operation or where gas-engined portable generators provide station power, but screened ignition cables typical of aircraft practice can increase system capacitance enough to compromise the operation of some recent capacitive-discharge ignition systems. (A diesel engine is one possible answer!) Other sources such as charging system, alternator, turn signals, fuel gauges, fan motors etc., can usually be dealt with by judicious capacitor by-passing, in-line chokes, and good bonding of engine and body components to the vehicle frame.[19] Similar treatment can also be effective on marine installations, but aircraft systems have their own specialized requirements which are well beyond the scope of this book.

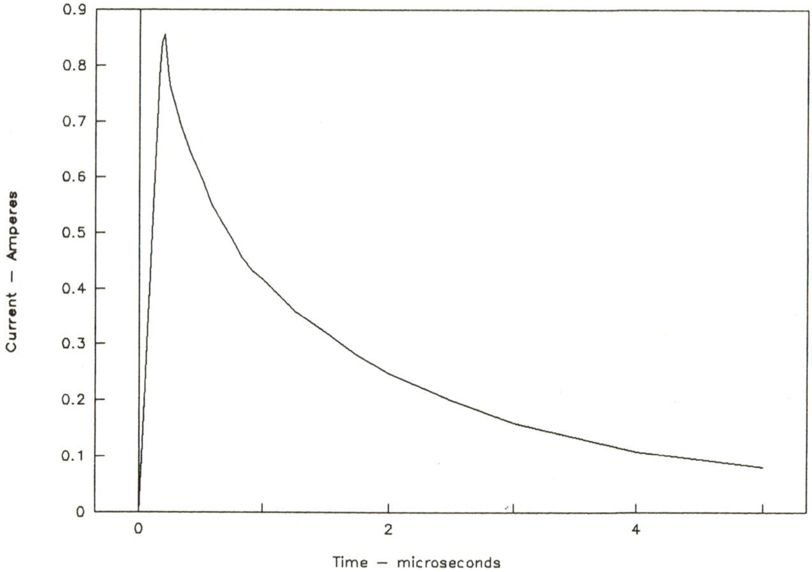

Figure 7.6 Typical spark plug current flow with a suppression resistor.

Arc tracking over the surfaces of line insulators or transformer (etc.) insulating bushings, and corona discharges, especially with extra-high-voltage lines (EHV, over 110 KV) cause most power line interference. Distribution lines at 33 KV and less, are more prone to connector arcing faults and re-radiation of conducted noise from other local sources. Permanent insulator failures such as cracked units, or temporary wet weather leakage, salt deposition near the sea,[16, 20] or bond failure in some H.V. pin type insulators[21] can all cause tracking. Bolted mechanical line splices can loosen and spark under load. Similar noise is radiated from overhead traction conductors, compounded by contact arcs and sparks at vehicle collectors.

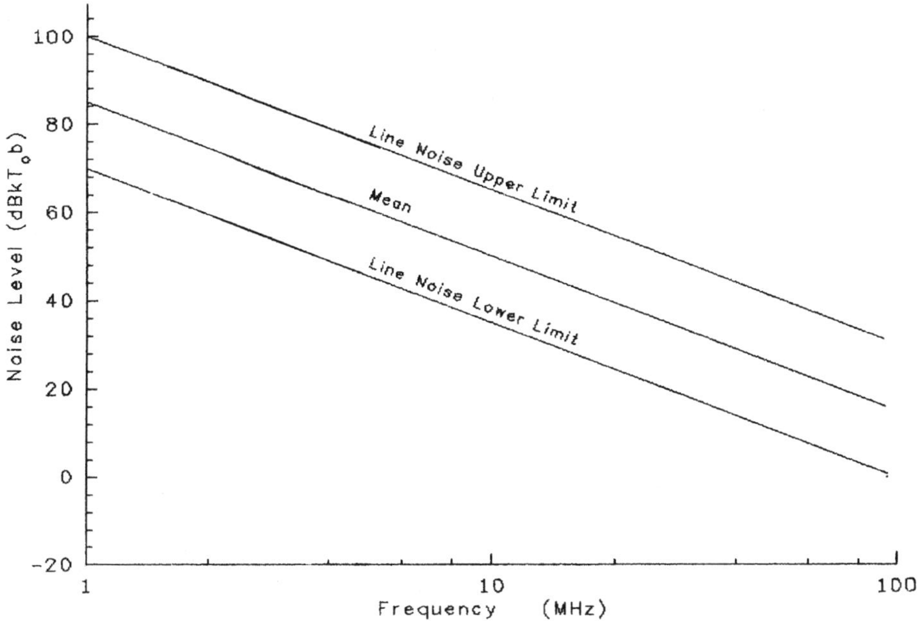

Figure 7.7 Typical overhead line noise, at ground level at the line location.

EHV line corona noise at low HF frequencies can increase up to 20 dB in rain and more in snow, and may be specially widespread when short sharp icicles form on conductors. These high potential points cause positive corona plumes with noise up to 25 dB above the lowest fair weather levels. These lowest levels are found when light winds dissipate space charges at any sharp discharge points.

Interference from large electric machinery may be less likely than expected, most larger drives using squirrel-cage AC induction motors with no moving contacts. Older variable-speed motors have brushes, but most installations since the 1980s use variable frequency AC, again with squirrel-cage motors. Switch operations do produce impulse noise, maybe a few times per day at shift end, maybe frequently.

Home and office equipment often has universal motors with commutators, rarely sees any maintenance, and can generate RF prolifically, radiating from wiring rarely screened by conduit or otherwise. Heavy industrial electrical wiring on the other hand, is quite likely to be screened by conduit, in US practice at least.

Domestic radio and television receivers are everywhere. Local oscillators are obvious emitters, an FCC standard having required less than 0.1μV/m at 1853 m. (1 n.m) or 400 pW emission through the antenna. [16] TV receivers can radiate line drive harmonics and 3.58 MHz color sub-carrier frequency (USA NTSC system). RF sources such as remote controls are usually only a problem at VHF. Personal computers were once important, but government regulation and greatly increased clock speeds have largely removed this source from the HF region.

Finally medical apparatus and industrial equipment such as RF heaters and arc-welders are quite stringently regulated. Under FCC rules, industrial scientific and medical (ISM) equipment is restricted in the HF region to two bands, 13.56 MHz $\pm$ 6.78 KHz and 27.12 MHz $\pm$ 160 KHz, except for certain industrial heating systems which may need to operate at a range of frequencies dependent on work piece dimensions. They are restricted by a complete ban in three protected frequency bands, (490 - 510 KHz, 2.17 - 2.194 MHz, and 8.354 - 8.374 MHz), with additional complex limits on permitted field strength. Studies of noise fields have shown maxima of 70 - 75 dBkT_0b at 5 MHz at 1000 feet radius (300 m.), with 50 - 60 dBkT_0b as a more typical level. [16] Any and all of these levels can unfortunately easily be made much worse by poor maintenance. Other rules define limits for industrial welders, but construction welding can present a much more difficult problem, fortunately usually limited in time and extent.

Site Noise Calculations

The graphical data of CCIR Report 258 provides the usual method of assessment of man-made noise levels. Four site types are recognized: Business, Residential, Rural and Quiet Rural, originally based on measurements made in the U.S.A., although more recent versions (258-4) also include data from the former U.S.S.R.

At 1 MHz these are 77, 73, 67 and 54 $dBkT_0b$, and with a 28 dB/decade negative slope, 49, 45, 39 and 26 dBkT0b respectively at 10 MHz. As a comparison, the report also shows galactic noise from 10 MHz to 150 MHz. Its level at 10 MHz is 29.5 $dBkT_0b$, with a somewhat lower slope of 24.5 dB/decade. Figure 7.10 includes the galactic or cosmic, business and quiet rural data.

However, it may also be useful, or at any rate of some interest, to look at a derived noise value for a given site. Three sources may be considered. Firstly, there is a general level which can be very approximately associated with a local center of population with all its businesses and residential noise generators. Secondly, proximity to a power line may associate with problem levels, and lastly, it is worth taking a look at the probable noise field of a traveled highway. Power line and highway noise decrement with distance is most likely to be of interest to the amateur operator, working typically on a fixed suburban site and seeking whatever respite may be found, even temporarily, from the onslaught.

Though correlation of noise with city size or population is not strong (some major cities have relatively few problem radiators, some small towns have many), Figure 7.8, adapted from a 5 MHz study,[22] says that a site selected at random in a city of population around 100,000 has about a 50% chance of man-made noise below 63 $dBkT_0b$, while a town of 10,000 presents a 70% chance of a noise level below 54 $dBkT_0b$. In a different approach, Figure 7.9 shows noise at 5 MHz for an industrial city of some 500,000 inhabitants, with its near suburbs at 2 to 4 miles, (5 houses/acre), far suburbs at 10 to 15 miles, (0.2 houses/acre), and at 30 to 40 miles, small farms with power transmission lines often to be seen in the locality. At 80 miles, overhead lines are seen only infrequently, usually 15 or 20 miles distant.[23] While this could represent a Midwestern US city, it is certainly not representative of the East Coast or of Europe! (It is in fact derived from Australian data.) Three atmospheric noise floor levels are shown.

The distance from some specific noise radiators to receiving antenna clearly affects final noise level, and though few sources make identifiable contributions at distances over 1 or 2 Km, all contribute to general background level. Far fields can be assumed, especially at higher frequencies, as direct radiation gives way to a Sommerfeld ground wave zone at about $d_{lim} = 2000/f^2$ m. (f MHz) with medium ground conductivity, [24] although measurement locations for power line and automobile noise have often approached near-field limits ($\lambda/2\pi$). [20, 25] This, and line elevation, cannot be ignored in applying such results as those of Fig. 7.7.

Very many studies have been made of ignition system interference levels, [16, 20] most of which study locations adjacent to highways, maybe 10 m. from the traffic stream. Levels expressed as $dBkT_0b$ (F_a) fall with increasing frequency across the HF band, although a resonance peak is found in the frequency spectrum at around 30 MHz. There is surprisingly little variation with traffic density; a few cars are almost as bad as a lot; but distance separation from a highway is still very useful. Noise generation is greater at higher speeds, but in compensation, faster vehicles remain close to the receiving antenna for proportionately shorter periods.

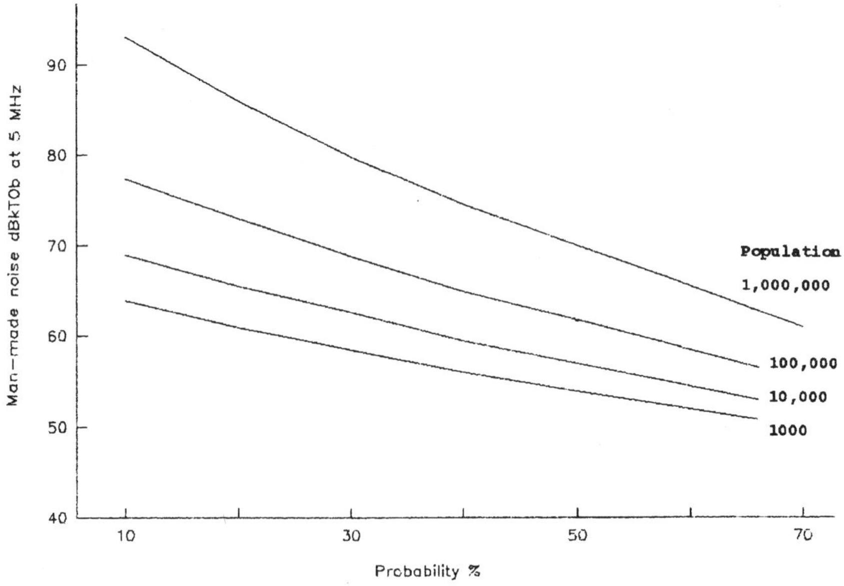

Figure 7.8 Noise level probability for a range of population centers.

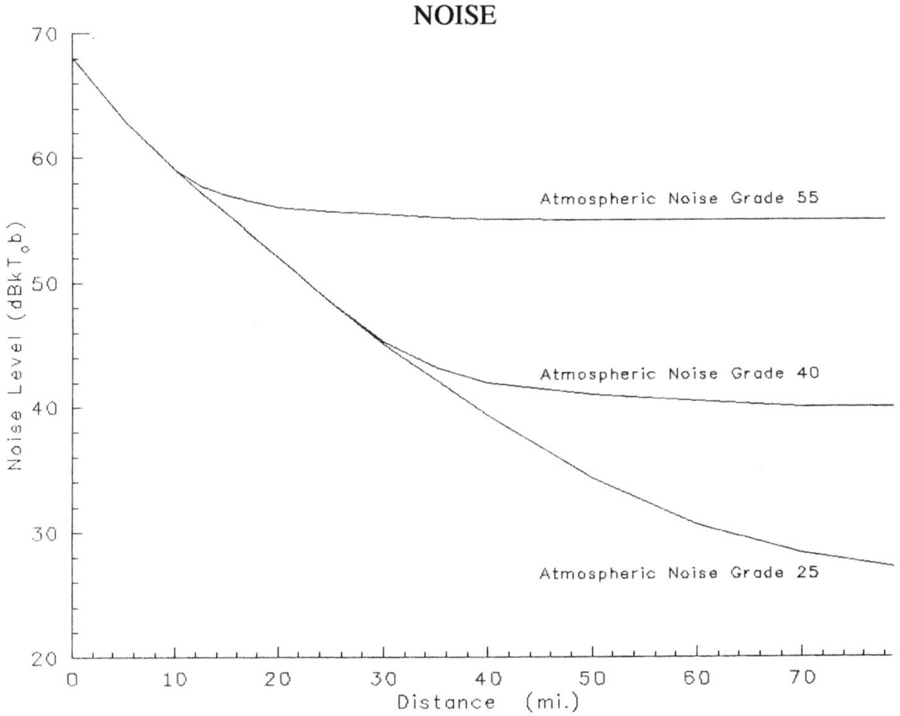

Figure 7.9 Noise level at 5 MHz vs. distance from a major population center.

Locally, radiated power line noise propagates as a ground wave, attenuated about 40 dB/decade of distance over real earth. Multiple sources, tower-to-tower spaced tracking insulators in rain, or a virtual infinitude of EHV line corona sources in a snowstorm, offer a complication; measurements close to a power line will favor radiation from line spans near the measuring location, while for a test site 1 Km away, many spans will be about 1000 m. distant, each to some extent acting as an equal independent source. Also, near field measurements may be made close to single conductors, while far fields balance radiation from all conductors. Finally, far field strength will tend to increase with line or antenna height, particularly for horizontal polarization and more surely at higher frequencies.[26]

Between these many causes, near fields typically decrease by some 20 dB/decade, and far fields probably about 30 dB/decade, reducing Figure 7.7 values by 30 dB or so at 1 Km perpendicular distance. [16, 18, 20] Measurements in the field are often obscured by unavoidable ambient noise, especially at higher HF band frequencies, but they may show the relative importance of particular sources.

Receiver Noise

Thermal electron agitation in receiver front-end circuits produces internal noise, and an antenna would produce similar thermal noise even if perfectly screened from all external sources. But antennas also capture all the noises that we already know about, atmospheric, cosmic and man-made. (We assume that all sources can be treated similarly to the random voltages of thermal noise, but this is only approximate, and can give incorrect results for such impulsive interference as automobile ignition.) Thermal noise in a passive circuit is given by the equation $w = 4kTb$ where w is power in watts, T is in $^{\circ}$K, b is bandwidth Hz and k is Boltzmann's constant. Alternatively, noise temperature $T = w/4kb$. If antenna noise is w_a, and external noise totals w_e, antenna "equivalent noise temperature" $T_A = (w_e + w_a)/4kb$. Considering the aerial as a perfectly matched generator feeding the receiver, it will send half its current forward to the receiver and half back to the source, so that an input noise power $w_i = k.T_A.b$ watts will add to the receiver noise power w_r. These are the powers that we will wish to compare. [27]

We aim for noise w_r to be somewhat below likely antenna noise input w_i, the noise floor often being specified in dBm (dB below 1 mW) for a given bandwidth. A typical recent review [28] shows a "noise floor" value of -123 dBm at 1.0 MHz and -132 dBm at 14 MHz, with a bandwidth of 500 Hz. This usage is often said to be inconvenient, as it varies according to tuning and to bandwidth. A noise factor f or *F* can therefore be defined, as an equivalent noise temperature, or as an equivalent dB figure, referred to a standard noise temperature, usually 290°K. Where $w_{A0} = k.T_0.b$, thermal noise power at the reference temperature,

$$f = (w_{A0} + w_r)/w_{A0} = 1 + w_r/w_{A0} \quad \text{or} \quad F = 10.\log_{10}(1 + w_r/w_{A0}) \text{ dB}$$

NOISE

This is usually academic below 30 MHz, with receiver noise level well below external noise, but low receiver noise can still pay off. If transmitters are also operating locally in HF bands, large extraneous RF voltages can appear at the receiver input, and overload is likely, with very probable intermodulation and spurious signals. If we can locate a minimum size receive antenna as far as possible from the transmitting antennas, we may avoid the overload; however, receiver noise will be more significant relative to the now attenuated incoming signals, and a low-noise front end design may well offer a distinct advantage.[5]

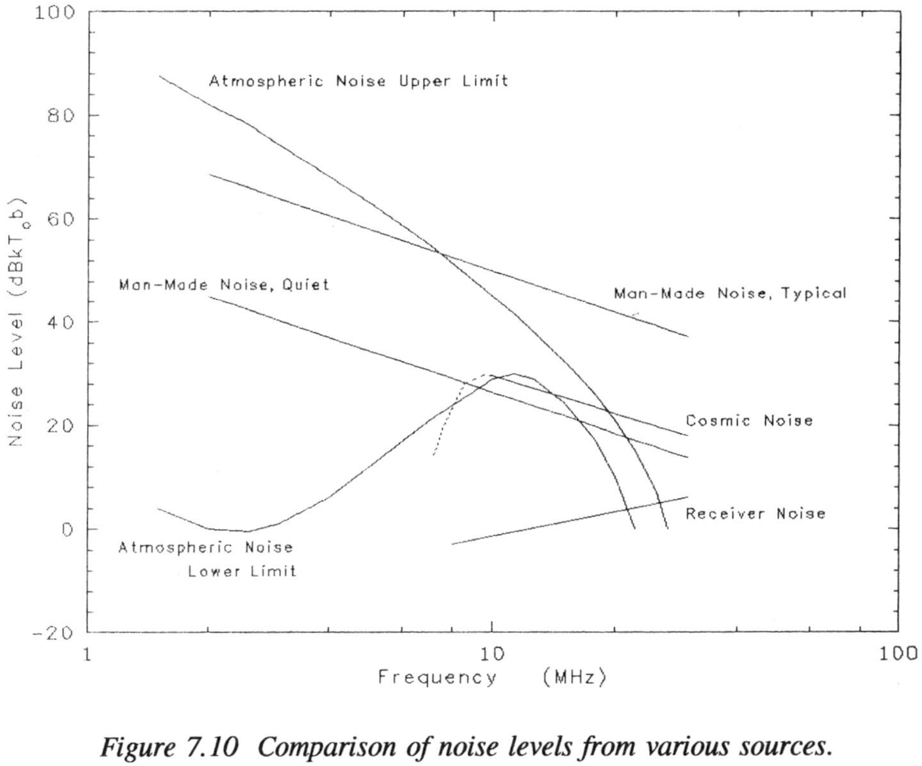

Figure 7.10 Comparison of noise levels from various sources.

Noise Level Comparison

In Figure 7.10 ranges of high frequency noise levels are shown for typical HF receiving installations. Atmospheric noise varies very widely with time and season, but at most world-wide locations it will fall within the limits shown. Receiver internal noise and cosmic sources are usually insignificant at HF, except in the very lowest noise environments, and even then, they are only important at the higher end of this frequency range. Clearly, man-made noise is usually controlled by very local conditions, but typical values are also included.

Notes and References

1 *Radio Spectrum Conservation*, JTAC 1952. In August 1919, the Inter-Allied Radio Conference proposed, presumably in some measure of desperation, that frequency bands should in future be allocated by country rather than by service, but that proposal was dismissed and never again lifted its head! The classic example of innovation running away from allocation was the development of short wave communication by amateurs in the 1920s, but JTAC cites several other instances, such as radar after WW2.

2 *Terrestrial electromagnetic environment*, Struzak, R.G.; see Rotkiewicz. This work covers radio noise and interference in considerable depth, somewhat academically. For an engineering overview see Horner, ref. 20

3 *Required signal-to-interference ratios for shortwave broadcasting*, Lane, G., Radio Science, v32, #5, p2091-2098, September-October 1997

4 Federal Communications Commission Regulations, USA, Part 97, and Ministry of Posts & Telecomms, Amateur (Sound) Licence A, U.K. 1971

5 Gustafson, also *Mutual Radio Interference in H.M. Ships*, Matthews, S.C. and Borrow, R.L.A., Proc. IEE v94, Pt. IIIa, 1947 p418-426

6 Volland, also World Meteorological Organization, 1953, see Kessler

7 Aiya

8 *Radio Noise from Thunderstorms*, Horner, F., see Saxton ed.

9 National Bureau of Standards Circulars 462 (1948) and 557 (1955)

10 CCIR Report 322. For maps and curves see also Braun and Saveskie.

11 ITU Radio Noise, also data embedded in such programs as VOACAP (*Voice of America* Coverage Area Prediction).
12 See for instance the case study for Howland Island in Chapter XI
13 *Radio Engineers' Handbook*, Terman, also ref. 19 below, and *Aerial Corona Interference*, Strafford, F.R.W., Wireless World, p441-3, Dec. 1948
14 Lovell covers early work, citing a series of IRE papers by Jansky & Reber. Meter-wave "Gun-Layer" Anti-Aircraft radars (GL1 Mk2) with rotary curtain arrays and chicken-wire mesh reflectors were still to be seen in military retirement (e.g. AACSTI (REME), Lydd, Kent) in England as late as 1950.
15 *Cosmic Radio Intensities in the VHF Band*, Cottony, H.V. & Johler, J.R., Proc. IRE, v40, p1053-1060, Sept. 1952
16 Skomal, also *Manmade Radio Noise*, Hagn, G.H.; see Volland. *Radiation from Receivers*, McDonald, G.J. & Thorn, D.A., Proc. IEE v94, Pt. IIIa p427-436, 1947
17 *Sparks & Flames, Ignition in Engines: An Historical Approach*, MacKeand, J.C.B., Tyndar Press, Montchanin DE, USA 1997
18 *Electrical Interference with Radio Reception*, Gill, A.A., Whitehead, S., Proc. IEE, v83 p345-394 1938
19 ARRL Handbook, (see Mobile and Portable Stations)
20 *Radio Noise Environment at High Frequencies*, Horner, F.; see Jones
21 *Power Transmission and Distribution Systems*, Enos, H.A., p14.112-14.121 Electrical Engineer's Handbook, (Electric Power), Pender, H., and Del Mar, W.A., eds., Wiley, New York 1949
22 *Man-Made Radio Noise*, Allen, E.W., see *The Radio Noise Spectrum*, ed. Menzel, D.H., Harvard University Press, Cambridge MA 1960
23 *Some Observations of Signal Strength of Skywaves and Noise Levels with Reference to Modes of Propagation*, Treharne, R.F.; see Jones
24 Picquenard, p166
25 Many noise measurements recorded in the field are quasi-peak levels. For conversion to r.m.s. values more convenient for site evaluation, Horner has suggested -10 dB; see also Rotkiewicz p228, 229 for CISPR definition.
26 Picquenard, p77-78 David & Voge, p92
27 David & Voge, p200-204
28 Product Reviews, e.g. QST February 2000.

CHAPTER VIII

ARRIVAL AT THE RECEIVER

Almost at the end of its travels, the minute remaining signal approaches the receiving antenna, together with all the potential interfering signals and noises to which antennas fall heir. The antenna makes a first cut, favoring some frequencies more than others, and the resulting collected ensemble travels on the transmission line, which is not loss-free, to the receiver, and it is entirely likely that some mismatch loss appears on the way. The receiver again favors some energy, hopefully including little besides the wanted signals, through its selectivity or capability of responding only within the required bandwidth, and finally the demodulator or signal processor may be able to distinguish and favor other known features of the wanted input, such as timing or encoding.

The Antenna's Contribution

It makes a lot of sense to take the very first opportunity to enhance the wanted signal at the expense of its competition, and we can do this most effectively by using all the knowledge about that signal that we can bring to bear.

Compass bearing, or direction, and ray angle, may be used as design data for the best antenna system to favor the wanted signal. However, unfortunately there are quite severe limitations on such designs, as the most directive antennas may be large and require considerable real estate, and must often be fixed in direction. The station location might also be a particularly important factor, although in most cases, antenna site is not a variable! Where such selection is possible, as in some portable operations, results can often be nothing short of spectacular. But while directional antennas can certainly help us greatly, the most highly directional antenna is not always the most desirable. If there is any need to scan a band for signals, then an omni-directional, or at any rate a relatively broad gain pattern may have quite an advantage. The time taken to rotate a large antenna to a new bearing can present enough of a problem that we decide to use, for example, a vertical instead of a multi-element yagi, or decide to invest in a separate antenna for an initial search mode. In general though, we do have some opportunity at this juncture to select or design an antenna system to bolster desired signals.

Design to favor predetermined ray angles is not usually practicable, but for portable work, signal enhancement due to sloping sites has been quite thoroughly studied. [1] Another exception is in short range Near Vertical Incidence Service (NVIS), where many antennas for low angle modes have near-overhead coverage so poor that quite distinct high angle designs are almost a necessity.

Low angle designs do discriminate against the higher elevation maxima of cosmic noise, horizon background levels being up to 15 dB lower. And finally, input from local noise sources can be minimized both by locating the antenna as far from them as possible and by taking advantage of any screening inherent in the site. While an antenna can help to discriminate directionally against interference from other stations, unfortunately it is not so clear that a similar discrimination can be effective against remote atmospheric noise sources. As we noted in an earlier chapter, many experiments have tended to show that for the sites of experimental location at least, no such advantage is necessarily available. [2]

Diversity

Propagation in HF bands is very variable, in the short term as in the long term. A signal can vary in strength and quality with time and space over periods as short as a few seconds, and distances as short as a few wavelengths, due to changing paths and arrival angles, and polarization too. We can take advantage of time variation with such modulation systems as Forward Error Correction, (FEC) which we will consider later, and in space by using several separate aerials. Two antenna feeds suitably combined can improve communication greatly, three even more so; not too often practical outside commercial point-to-point service, but an extremely effective way to obtain low error rates. Improvement is measured as diversity gain, the increase needed in signal level to get the same results as the diversity scheme being considered. Figure 8.1 is for two antennas spaced along a line towards the transmitter, zero spacing representing a single antenna and no diversity. The diversity gain if we require 99.9% usable signal time is about 12.5 dB for six wavelengths spacing, and at 99% it is about 7.5 dB, but if our need is only for 93% usable signal then the diversity gain falls to a mere 4 dB. It appears therefore that with the modest accuracy needs of amateur radio, diversity is somewhat less of a bargain than it is for the professional.[3]

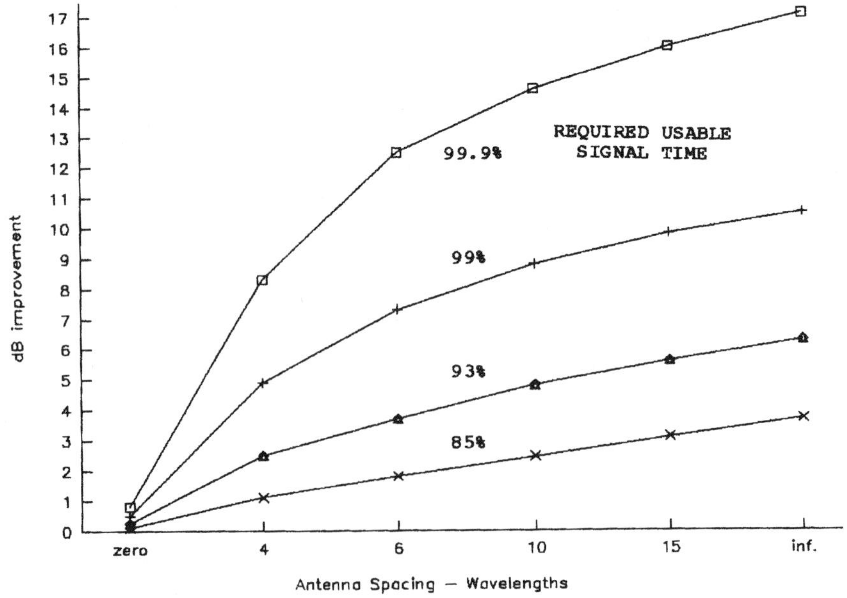

Figure 8.1 Signal Improvement due to Diversity

Modulation and Bandwidth

Early in radio engineering the maxim "*The wider you open the window, the more dirt blows in*" was a recognized truism. [4] If receiver bandwidth is increased, as soon as it is greater than the signal requires, noise input continues to accrue, while no more signal power is to be found. There are many reasons for using wider bandwidth; equipment limitations may dictate it, frequency instability at either end of the link may require it, ancillary operations such as panoramic band scanners may need it, ease of tuning may advise it; but there is a price to pay.

To take best advantage of a given circuit, we need to use all the information available to us. Amplitude modulated (AM) HF broadcasting offers a simple example. Listening to BBC or Radio Netherlands news, modulation frequencies above 2800 Hz add little to intelligibility, and we can select a 3 KHz bandwidth, a wider band merely decreasing signal to noise ratio. If a music program follows, then a 6 KHz or even wider band is indicated, but unless signals are quite good, increased noise level and interference from adjacent channels may result in much less satisfactory reception. [5] (Single sideband operation is assumed; direct amplitude modulation (A3) uses both RF sidebands and so doubles band usage.)

Frequency modulation (FM) presents a more complex picture, with a threshold of improvement due to the inherent capture effect. Usually this is invoked to explain the almost complete domination of a strong FM signal over even a slightly weaker close or co-channel signal. The same reasoning can also be applied to signal-noise, although complete capture from noise only occurs at a fairly high signal-noise ratio, in the order of 10 dB. The statistical qualities of noise ensure that in the threshold region, high noise peaks will momentarily capture the channel from a stronger signal. When FM is good it is very good, and when it is bad it is very bad, but clearly it is not at its best as a weak signal mode.

The digital modes provide the most striking examples of the use of inherent information within a signal stream to enhance noise discrimination capabilities. It is easy to see that if we know that a signal will only change state, from on to off, or from mark to space, at certain predetermined times, then we can discriminate

Mode of Modulation [6] (Baud rate - Digital)	System Bandwidth Hz	Poor SNR dB	Mid SNR dB	Good SNR dB
PSK (31 Baud) [7]	31	23	26	29
CW (8 Baud)	250	21	31	41
RTTY (45 Baud) [8]	500	37	45	53
TOR (100 Baud)	500	30	40	50
SSB	2600	46	56	66
DSB AM	6000	52	62	72
SWL Broadcast [9]	9000	65	75	85
FM	15000	67	77	87

Table 8.1 Signal-Noise Ratios referred to 1 Hz Bandwidth

The normalized signal-to-noise ratios suggested in this table [6, 7, 8, 9] may be converted to SNR for the bandwidth b (Hz) in use, subtracting 10 $\log_{10}$ (b) given in Table 8.2 below.

Bandwidth Hz	10	32	100	320	1000	3200	10000	32000
Correction dB	10	15	20	25	30	35	40	45

Table 8.2 Correction to Signal-Noise Ratios for Actual Bandwidth

against state changes that don't follow the rules, assume they are due to noise and ignore them. The oldest example is Morse code reception by ear, the human brain using its remarkable pattern recognition capability to reject the less likely signal combinations in favor of more probable letters and words. A development known as coherent CW has semi-automatic code generation, reversal timing depending on a common time source, maybe WWVL or another standard frequency station.

Radio-teleprinting systems such as Baudot RTTY have spawned their own crop of improvements, from the mechanical rangefinders of the Model 19 and 28 machines through sophisticated hardware demodulation devices. More recently a spate of software based systems has emulated and improved on earlier RTTY with such modes as AMTOR, Pactor, G-Tor, HF Packet and Clover. [10] These employ ARQ operation, in which an automated to-and-fro transmit-receive handshaking protocol sends an acknowledgement based on a pattern analysis of a just-received defined group of the incoming signal, thereby either requesting further copy or a repeat of the previous group. The high level of HF noise dictates a fairly short group, while quiet VHF circuits benefit from the much longer "packet" group. Copy percentages are also enhanced by using internal code structure information.

Forward-error correction (FEC) does not use handshaking, all transmitted code being repeated in a form of time-diversity. Two (or even more) received versions are compared, and code design allows error detection. Other recent developments such as PSK31 also correct many errors before the decoded signal is displayed. This protocol was designed by its author for keyboard to keyboard conversational working, with outgoing signals formatted to minimize bandwidth. Unlike many machine telegraphy derived systems, its sophisticated coding takes a leaf from Morse code's book. Variable length symbols economize on bandwidth for a given language and text speed; frequent letters use short characters, less frequent letters and operators use long ones. [11, 12] Also in use today are the previously almost forgotten Hellschreiber, [13] (or Feld-Hell), and computer-based variations on this theme such as MT-Hell [10] and the multi-tone systems exemplified by Piccolo. Slow-scan television needs its own analysis of signal-noise effects, beyond our present scope, and now narrow band digital telephony has appeared on the scene, potentially bringing many digital processing advantages to voice modes. [14]

Notes and References

1 Moxon, p138

2 CCIR Report 322

3 JTAC, 1964 p82-88, Braun, p118-122, Goodman, p274-8 and p348-9

4 I can't find the source, but believe it was P.P. Eckersley c. 1930, quoted in a long lost *Wireless World*, probably by "Diallist" or "Free Grid".

5 Clearly there is a large personal preference element in the acceptability of music reception. Listeners are far from agreement on audio reproduction, some types of music apparently being very much more, or less, demanding than others. See ref 9 for a different view!

6 ARRL Handbook, RSGB Handbook and Teleprinter Handbook, Braun, CCIR 339-5, ITU *Red Book*. *Radio Antenna Engineering*, Laport E.A., McGraw-Hill NY 1952 cites figures in Appendix VII for many services, stated to include both atmospheric noise and interference.

7 Private communication (1999), Martinez, P., G3PLX

8 Goodman, Fig. 5-18 cites Hagn, G., 1988 with FSK error rates vs. SNR

9 Acceptable broadcasting signal-interference ratios are much higher than signal-noise ratios, generally being around 50 dB. *Required signal-to-interference ratios for shortwave broadcasting*, Lane, G., Radio Science, v32, #5, p2091-2098, September-October 1997.

10 ARRL Handbook has a good introduction, but the Internet is today's best source of information on these rapidly developing technologies.

11 *PSK31: A New Radio-Teletype Mode*, Martinez, P., RadCom, V.74, #12 p14-16 Dec 1998, and V.75, #1 p26-26 Jan 1999 RSGB, London

12 *PSK31 - Has RTTY's Replacement Arrived?*, Ford, S., QST, V.83, #5 p41-44 May 1999, ARRL, Newington CT.

13 *Modern Telegraph Systems and Equipment*, Perkins, W.T., George Newnes Ltd., London 1946. *Hell* provides an ultra-low definition fax-like view of the signal, which can be read by eye through very heavy noise and interference.

14 *Practical HF Digital Voice*, Brain, C. & Talbot, A., QEX, p3-8, May-June 2000, ARRL, Newington CT.

CHAPTER IX

FINAL ASSEMBLY

Now, at last, if a projected path proves to be available, all the losses, and the very few gains, can be combined with the transmitted power into a single number which will describe the strength of the received signal; a similar number can describe the total strength of the noise at the receiver; and then we can finally derive the signal-to-noise ratio for the path, time and equipment of interest, and see if it meets our needs. It's no trivial task, and before easy access to digital computers it was a long laborious business. Some of the first widespread help appeared in the early 1940s as curves in radio engineering handbooks,[1] although these were often illustrations rather than aids to a solution. By the mid 1940s contour charts of critical frequencies were being produced, as well as transparency maps for operational use.[2] However, world-wide absorption data to allow calculation of received field strength was not yet necessarily accepted as valid.[3] In 1963 numerical mapping methods and derived maps for MUF calculations were widely available, with some indications of HF field strength.[4] By the 1970s computerized HF propagation prediction services were able to provide rather complete propagation and reliability calculations on a quasi-commercial basis,[5] and through the last 20 years many desk-top computer programs have come into commercial, military and non-professional use for all the basic calculations.[6] Examples are often preferable to explanations, so three very different studies are offered to illustrate some of the many angles we have already seen, and to show a few of the different situations in which propagation calculations can be applied.

The first one looks at apparatus and conditions when Guglielmo Marconi first tried to bridge the Atlantic Ocean with his radio signals in December 1901. Surely that was a long wave operation? But, no standard calculation for wavelengths on which he was thought to be working shows even the slightest chance of readable signals from his transmitter in Poldhu, Cornwall, ever appearing in the telephone receiver at St. John's, Newfoundland. And his later commercially successful efforts bore little technical relation to that first signal reception. [7] As a result, many very competent engineers still strongly believe that he deluded himself. Detection of the three crackling dots of a spark transmitted Morse "S" amid the atmospheric background noise is not easy to support. Tight commercial secrecy around the initial trials, and the subsequent publicity impelling his company to continue expensive and often fruitless tests, have in themselves been almost enough to ensure a very skeptical approach. However, there's another possibility. The big spark station in England will have put out quite a large amount of radio frequency energy on wavelengths other than the principal one. Maybe enough such energy was produced, and then radiated with good enough propagation on a wide range of much higher frequencies in today's HF bands, that the three dots really did cross the ocean, and really were heard on that cold windswept headland in distant Newfoundland.

In the second case, Amelia Earhart disappeared in July 1937 in the middle of the Pacific, near the end of a round-the-world flight. [8] Since that time rumor and journalistic endeavor have placed her in Japan as a prisoner, in New Jersey as a housewife under an assumed name, and on several island groups near and far from her intended immediate destination. However, the use of well established propagation data, together with original Coast Guard radio communication logs and detailed contemporary information on aircraft and shipboard radio installations has enabled us to pin-point her actual locations shortly before final ditching within a surprisingly small area.

Finally, three contacts between amateur radio stations are examined using the computer program *SNAPmax*. The first is a long-distance, almost antipodal situation, the second is at a transatlantic medium range, and the third is very local and non-ionospheric. The *SNAPmax* propagation software program is available as "freeware" for non-commercial use, is based on the methodology described in this book, and is intended to help develop a picture of how H.F. circuits and their losses depend on the rich background which has been introduced here.

Notes and References

1 *Reference Data for Radio Engineers*, Federal Telephone and Radio Corporation, New York 1943 and its British associate, Standard Telephones and Cables Ltd., London.

2 Bennington, p90-100, see also Tremellen & Cox

3 Bennington, p122

4 CRPL Handbook, also see ref. 1, 1949 edition

5 Technical Report I.T.S.A. #1, Environmental Science Services Administration, Boulder, CO 1966 cited e.g. in *The Practical Prediction of Lowest Useful Frequencies*, Haydon, G.W., see Jones, also Saveskie, p84

6 Goodman, p370-2, Jacobs, Cohen & Rose, p6-1 through 6-28

7 Many references for Chapter X also refer to Marconi's later work, much of it in Canada at Glace Bay, Nova Scotia. *Whisper in the Air, Marconi - The Canada Years 1902 - 1946*, Mary K. McLeod, Lancelot Press, Hantsport NS 1992 is also a useful, mainly non-technical, introduction to this later period.

8 Many books and many more articles have been written on Amelia Earhart and her flights, especially her last. Some have resulted from serious study while others are unfortunately no more than journalistic speculation. Two excellent references are *Sound of Wings* by Mary Lovell (ref. 3 Ch. XI), and *Amelia Earhart - The Mystery Solved* by Elgen M. Long & Marie K. Long, Simon & Schuster NY 1999.

CHAPTER X

THE FIRST TRANSATLANTIC HF ?

Only six short years after Marconi's first public demonstrations of wireless telegraphy on Salisbury Plain, he was to attempt the impossible. Working ranges of W/T equipment were still hardly more than 300 km. at most, and so the transmission of wireless messages across the thousands of miles of the Atlantic Ocean could surely be no more than a wild dream. But to implement that dream, Marconi built in 1901 at Poldhu, Cornwall, almost at the extreme western tip of England, the most powerful spark transmitter yet envisaged, powered by an oil-engined generator, when predecessors had used batteries. The station was completed by a particularly elaborate aerial system, supported by a spectacular circle of 65 m. wooden masts, while across the Atlantic, in Wellfleet on Cape Cod, Massachusetts, he erected a similar antenna to receive the Poldhu signals.

North Atlantic weather is never an easy colleague, and for Marconi it was less than co-operative. In September the worst gale in living memory blew down the tall Poldhu masts, and the Cape Cod array followed suit shortly thereafter on November 26th. His tenacity of purpose accepted no setbacks. A temporary aerial was strung at Poldhu between two re-erected poles, now less than 50 m. high, and he sailed with George S. Kemp, his assistant, not to Cape Cod, but to St. John's in Newfoundland, rather closer to Cornwall. There, about midday on a windy December 12th. 1901, with the simplest untuned receiver and a kite-flown wire antenna he listened for the famous Morse Code "S". In those earliest days, with no well developed method of measuring frequency, the wavelength was unknown; the impact of daylight on radio transmission was only discovered in the following year; certainly the practical effect of the ionosphere in all its ramifications was yet unsuspected, and both sites were beset with equipment failures. [1, 2]

His claims of success, not too surprisingly, met with very mixed responses. On the one hand were those who eagerly looked forward to breaking the effective trans-oceanic communication monopoly of the cable companies, and on the other hand was the scientific community, whose calculations appeared to demonstrate quite conclusively the impossibility of the task. It may be even more surprising that modern calculations have only exacerbated that aura of doubt, [3] showing that with the power available to him, and the wavelength on which it was later believed that he had been operating, no signals were in the least likely to be heard. However, M.A. Cross and the author proposed, [4] and also found that a suggestion had been made years before by G.R.M. Garratt of the Science Museum in London to the same effect, [3, 5] that the signal heard might have been the integrated effect of transmission, over a wide band, of high frequencies produced by the open spark gap somewhere below the Maximum Usable Frequency (MUF), and propagated over a relatively low attenuation path at a time in the solar cycle when ionospheric absorption was at the extreme low point of Cycle 13/14.

To test this theory, we constructed a combined hardware and software model of path and propagation, conditions and equipment, repeating the process at a hundred or so frequencies from 3 KHz to 16 MHz, each representing the center of a small frequency band, from the lowest effective frequency for atmospheric noise to well above the calculated MUF. The antennas were simulated by software models using the method of moments, and ray angles were determined from path calculations assuming F-layer propagation. The transmitter was tested as a physical model of the "jigger" or tank and coupling coils and tuning capacitor, terminated in components representing transmitting antenna impedance, as we strongly felt that software models could not easily capture the essential rich mix of inductance and capacitance, intended and stray. And the strays are key to our proposal of significant HF power output. [6] Propagation was handled in software at each frequency. The whole range of results was then assembled to obtain three power density spectra; transmitted power from Poldhu, received signal power at St. John's and received atmospheric noise power at St. John's. These were reviewed in the light of known statistical variances in noise and propagation, and of the properties of the human ear listening to impulsive signals, to produce a final estimate of a likely effective signal-noise ratio at the relevant times of day.

SIGNALS, NOISE AND PROPAGATION

Poldhu

The transmitter aerial in Cornwall was a temporary structure.[1, 2] It replaced the original 65 meter high circle of wooden masts, erected under the oversight of seamen still very familiar with wooden ships and spars, and was apparently seen as an appropriate and economical way to avoid unknown effects of conducting iron structures. Here good engineering design maybe eluded the company, but hindsight is too easy. The substitute was simpler; two masts, each some 45 to 48 meters high, supported fifty, maybe fifty four, wires fanning out from a common base to a triatic stay rope 60 m. long, between them. A simplified model with rather fewer wires appeared to be adequate for a satisfactory analysis.[4, 7]

The Poldhu transmitter was not at all typical of Marconi's equipment even a few years later.[8] Some of its components were well recorded, while others have been carefully extrapolated from the available data, so that it is possible to reconstruct the system with a fair degree of confidence. It had an unquenched open spark gap with a tank condenser of known size, coupled to the aerial by two coils known as the "jigger". We followed Thackeray[7] in his analysis of this component.

Wavelength was not recorded, and surely not power distribution with frequency. But with a short spark gap so that rise time is rather less than the period at any frequency to be generated, power-frequency distribution can depend only on tuning components and antenna, stray high frequency resonances being excited as readily as those for lower frequencies. Compare this with lightning discharge frequency spectra, which depend on spark-length,[9] and the earliest millimeter-wave generators with extremely short spark gaps. So we see the spark as a white noise generator coupled to jigger and antenna, and studies of overhead line noise from localized sources, corrected for line losses also support this view.[10] Contrariwise, it has been pointed out that Fourier analysis of pulse sequences, and measurements on simple spark-capacitor systems can show 1/f proportionality.[6] However, to satisfy this case, the pulses must be flat-topped, which spark impulses show little sign of being, and in one reported experiment which has been analysed, a 1/f decay appeared to be due to a circuit configuration which placed an RC filter of capacitor and spark source resistance in the measurement path.

The spark has therefore been considered as a fast switch, any and all resonances being excited by the same voltage step, and all decaying from the same initial voltage, independently of frequency. So we excited our jigger with spark-generated energy from a small buzzer contact, battery driven to avoid spurious power system couplings. The output was loaded at each test frequency with components representing input impedance of the transmitting antenna, and we measured jigger input and output voltages at each frequency, so that the ratio thus obtained could represent the energy transfer to the aerial.[4]

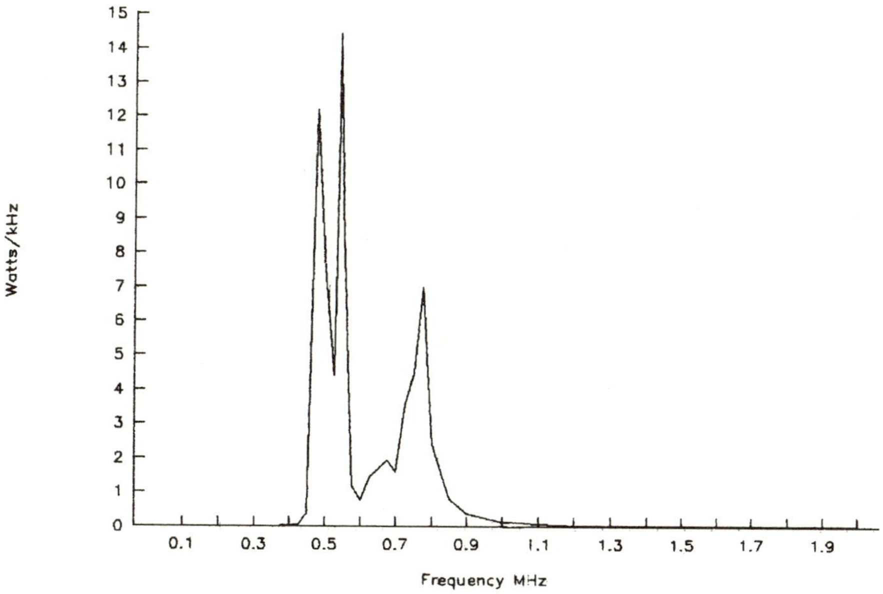

Figure 10.1 Power radiated from Poldhu at low frequencies.

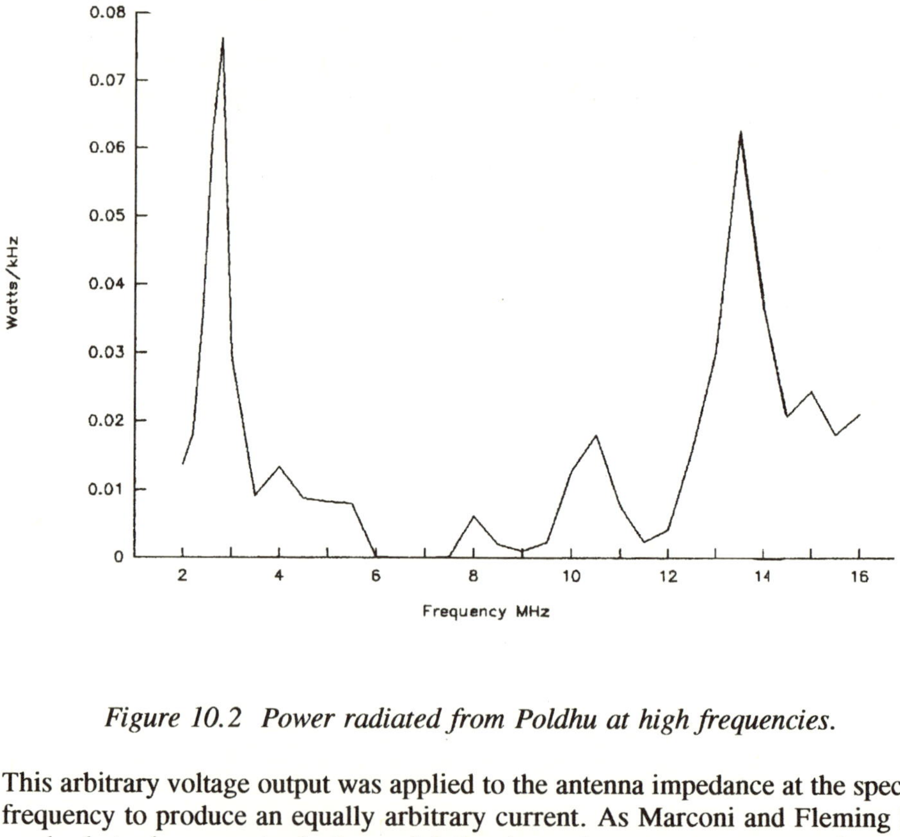

Figure 10.2 Power radiated from Poldhu at high frequencies.

This arbitrary voltage output was applied to the antenna impedance at the specific frequency to produce an equally arbitrary current. As Marconi and Fleming had used a hot-wire ammeter in the aerial circuit, each current was summed over its associated bandwidth, and then all the bands were r.m.s. summed to give a total output current. The arbitrary reference level for all fifty output measurements from 100 KHz to 16 MHz was then adjusted to give 17.5 A. total aerial current as recorded by Fleming.[11] The adjusted voltages, and the aerial radiation resistance at each center frequency, were used to calculate radiated power and power density for the represented band, as shown in Figures 10.1 and 10.2, and later CW tests showed essentially similar results. The theory of operation of such a transmitter expects a "double-humped" output power curve, and indeed the two humps do appear, at 475 KHz and 540 KHz, not far from the traditional marine frequencies, which in their turn derived from the first spark transmitter operations of the the early 1900s. Summation about the humps shows an "on-frequency" radiated power of about 950 W, and total radiated power of about 1800 watts. There is a smaller peak at about 800 KHz, and Figure 10.2 shows some 140 W being radiated in total in the high frequency region between 2 and 16 MHz. Power at frequencies over 16 MHz, well above the MUF, was irrelevant to the transatlantic path, and insignificant in terms of total power produced.

St. John's, Newfoundland

The Signal Hill receiving system with its long kite-borne wire, blown away more than once from the exposed clifftop, [2, 12] was also modeled. [13] As at Poldhu, the antenna was situated on a headland overlooking seawater in the intended direction of transmission, clearly a very favorable factor. Then, using appropriate ray angles, the antenna was analysed for terminal impedance and dBi gain. [13]

Figure 10.3 "Italian Navy coherer" receiver at St John's, Newfoundland.

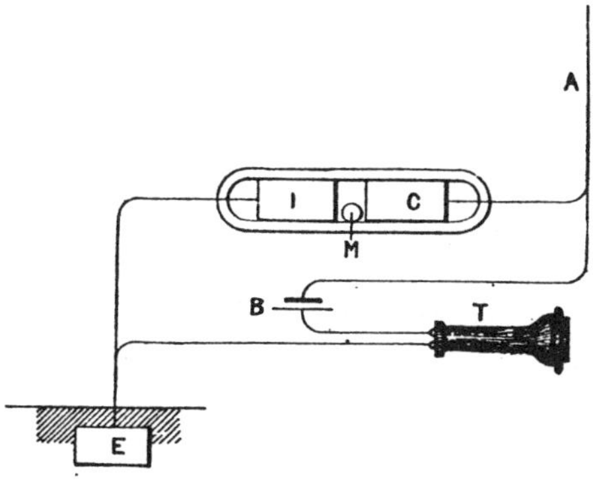

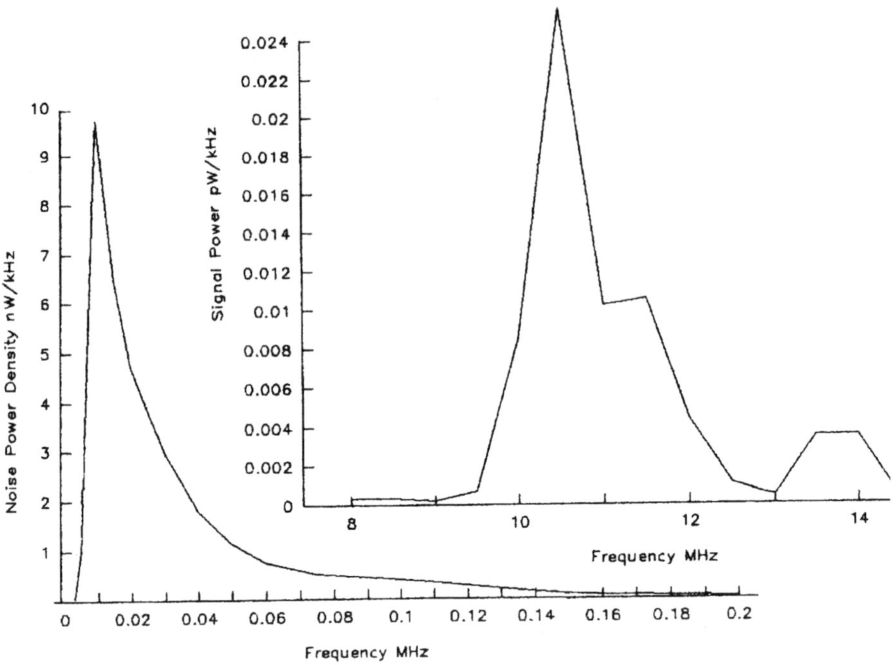

Figure 10.4 Signal power from Poldhu and atmospheric noise power received at St. John's, Newfoundland.

The aerial had dictated the receiver selection. With Marconi's preferred tuned or "syntonized" receivers, antenna capacitance was one element of the tuning system, and while this worked well enough with a fixed installation, the unstable wind-blown wire forced him to use "self-restoring coherers" as detectors, and therefore incidentally to rely on aural reception rather than the inker systems preferred at the time. Two such devices were used at St John's for successful reception: a Marconi design using "loose carbon filings" and the "Italian Navy" mercury-iron coherer.[2, 14] His usual receivers used conventional coherers with metal filings and a reset tapper, but these apparently gave only doubtful results in the tests.[2] Self-restoring coherers, maybe best described as biased low-resistance (1000 ohm) thin-film-breakdown devices, were usually connected directly from aerial to ground, in parallel with a telephone receiver and series dry cell,[15] although a transforming "jigger" may have been used at Signal Hill.[1] Not reliable enough for commercial application in 1901, nor readily able to operate code recorders, they were said to be very sensitive for experimental work. One contemporary investigation of coherers found them sensitive to radio frequency power levels as low as 10 nW,[16] and sub-microwatt signal levels were certainly audible with telephone receivers of the types which appear likely to have been used.[17]

Man-made electrical noise would not be very evident in Newfoundland for some years to come, and galactic noise could be neglected given the relatively high F-layer cut-off frequency. Atmospheric noise at St. John's controlled at this season although it was also fortuitously low,[18] but all was grist to this untuned system's mill and summation was extended down to 3 KHz. Almost certainly there were no local thunderstorms, and most probably few, if any, closer than the major South American and Central African storm centers.[19] The background noise, from many distant tropical thunderstorms, would therefore have propagated in ionospheric wave guide mode as a dull integrated roar sounding more like random thermal noise than the familiar crash of individual lightning strokes typical of local LF storm noise. Figure 10.4 shows received noise power density, from noise grade data summed from 0.003 MHz to 0.5 MHz. Mean total power is 210 nW, and 6 dB less at 52 nW for 10% of the time. Its density peaks about 10 KHz and 70% of the effective noise power arrives at frequencies below 50 KHz.

Propagation

Losses on the oceanic path were calculated for signals from 2 - 16 MHz by methods described in the earlier chapters of this book, and it was assumed that Marconi encountered neither ionospheric storms nor auroral absorption! Sunspots were at the very lowest point of the cycle in December 1901, with sunspot numbers of 2 or 3, solar flux probably in the range 66 to 76, and maximum usable frequency most likely hovering around 10 MHz. At lower HF frequencies multiple paths probably existed, but at higher frequencies near the MUF, where absorption was lowest, single-hop F-layer propagation was predicted, attenuation being far and away lower than for the MF fundamental in the 500 KHz region. So we decided that Poldhu's signal, if it was heard at all, was most likely received between 2 and 16 MHz. That range was therefore integrated, combining radiated power with absorption and other losses, and gain of the receiving antenna at each center frequency. The calculated frequency distribution of the signal power is shown in Figure 10.4, with a total of about 50 pW. Fading is also relevant at HF, and while signals lose strength in most of the fading range, they are typically about 5 - 7 dB above the mean during some 10% of the time.[20]

Received noise level is not immune from the vicissitudes of propagation, and must be adjusted for a path attenuation appropriate to the low level of solar activity. Transmission was almost entirely at very low frequencies (VLF), and with low solar flux, VLF attenuation (unlike LF) increases.[21] At an approximate distance of 7500 km, with an average sunspot number of 3 for the previous year, noise from major tropical storm sources would be over 4 dB lower than levels predicted from noise grade data alone, for a final mean received noise power of 82 nW.

Signal-Noise Comparison

As we saw in Chapter VI, we cannot directly compare random thermal noise with impulsive signals, and the Poldhu signals were certainly very impulsive, while the noise, most remarkably, was not. Fleming estimated that the Poldhu transmitter produced 21 sparks/second, that is, at intervals of about 0.05 sec,[7, 11] with a spark duration of 5 μsec, a duty cycle of about 1/10,000. With a CW signal, the signal-to-noise ratio would simply be the difference in received power levels, but with a 1/10,000 duty cycle, the same average transmitter power, and a peak pulse duration of say 1 second, or even 1 minute, the received signal during that period would be enhanced by 40 dB. It's not at all obvious what effect short pulses might have on the human ear, however. Clearly there will be a real gain, as Garratt and Ratcliffe both pointed out,[3, 5] and it will be between 0 and 40 dB. A set of audio experiments showed a simple relationship, which is at least in part explainable by Aiya's work[9] on pulse interference in tropical short-wave broadcasting.

But first we need comparable noise characteristics. On average, at any given time some 2000 thunderstorms are in progress world-wide, making about 100 lightning discharges per second,[22] each with three or four radiating impulses,[9, 23] for a total of around 400 impulses/sec. In the location and at the time in question, noise originated almost exclusively in tropical storms of two extremely remote but very large areas of Africa and South America, with most propagation at very low frequencies. December storm area maps[19] show maybe 60% of the world's active storms as likely contributors, for an estimated effective total of 250 impulses/sec. Noise "crossing rate" studies, although not exactly comparable, also show a peak in this mid-audio range, as does informal winter LF listening in a period with no local storms. There is therefore a useful aural distinction between the "middle C" of the far distant storms, dulled also by the narrow effective bandwidth of VLF ionospheric waveguide transmission,[24] and the clatter of the sharply impulsive wide-band Poldhu signal with pulse repetition frequency over three octaves lower.

Aural signal-to-noise ratio is much affected by auditory masking. The large body of information on the subject[25] says clearly that higher frequencies are relatively ineffective at masking lower, but hardly addresses signal energy. An experiment was set up comparing noise-plus-signal with noise, to find minimum discernible pulse power in a telephone receiver as a transducer, with noise first unpeaked and then broadly peaked at 250 Hz with a low Q (15) LC filter. The signal was a series of narrow pulses RC differentiated with a 7 μs time constant on the rise of a square wave, power level being calculated from the duty cycle and DC input to the generator. Recorded noise was simultaneously played back into the same transducer, power level being determined by loading an incandescent lamp with noise or DC. At a 60 Hz repetition rate with no noise peaking, pulses were heard down to 7 dB below noise level, and with 250 Hz peaking, down to 19.8 dB below the noise. A later experiment demonstrated that pulses at frequencies from 20 Hz to about 200 Hz were heard through the noise at a well-defined amplitude independent of their frequency. Thus audibility of a pulse stream in this noise background was dependent only on the energy content of each individual pulse, a signal-to-noise power ratio of 24.4 dB being demonstrated at Poldhu's 21 Hz repetition rate, although it appears from the literature that actual values may be quite dependent on the ear, the transducer and test power level. A subjective impression from the experiment should also be noted, that signal recognition was much enhanced by the regularity of the pulse stream, even at 20 Hz.

An explanation may be hazarded. The ear hears impulses as a peak-picking device with charge and discharge time constants of about 10 ms and 500 ms., the full strength of any sound being judged in about 200 ms. In a 200 ms period, 50 noise impulses would be received, against only 4 Poldhu impulses. Aiya has noted[9] that at repetition frequencies as low as 20 impulses/sec, the ear responds almost fully to peak pulse amplitude in assessing loudness, and higher impulse frequencies do not increase perceived loudness. Therefore 45 of the 50 noise impulses in any 200 ms period are ineffectual compared to signal impulses as 5 alone will produce a full aural response. Therefore in calculating the signal-to-noise power ratio, 5/50ths or 10 dB can be attributed to peak-picking. The ear's pitch selectivity, its ability to follow a double bass's thread in the midst of wind, strings and brass is probably important too. This may well be the *regularity* enhancement mentioned above, and the remaining 10 dB of the experimental result may be so attributed, but it remains as a possible and interesting future investigation.

Conclusions

Some 50 pW of wide-band HF signal was received at St. John's. Integrated LF noise power was 210 nW, reduced by VLF transmission loss to 82 nW, for a "power signal-noise ratio" of -32 dB, defined on the basis of average received power levels. HF fade range has a typical upper bound of +6 dB for 10% of the time, while the lower decile of the integrated noise is 6 dB below the mean for a different 10% of the time, varying randomly with respect to signal fading. Power signal-noise ratios therefore reached -20 dB or better for around 1% of any hour. But due to the peak-picking phenomenon, and maybe aural selectivity, pulsed signals like that of the Poldhu transmitter can be distinguished at levels down to -20 to -24 dB power signal-to-noise ratio.

THE FIRST TRANSATLANTIC HF ?

Daytime HF propagation calculations showed that the largest signal strengths would have occurred around the claimed reception period of 12.30 to 2.30 pm local time, [1] signal-to-noise ratio decreasing rapidly before and after those hours.

In the light of this analysis, it looks as if Marconi may very well have heard in Newfoundland in December 1901, the integrated effect of 5 to 15 MHz high frequency sky-wave signals propagated across the Atlantic Ocean from Poldhu in England in a band several MHz wide, just below the maximum usable frequency.

Notes and References

1 *A History of the Marconi Company*, Baker, W.J., Methuen, London 1970
Marconi and Wireless, Vyvyan, R.N., EP Publishing, Wakefield, UK 1974
Marconi's Atlantic Leap, Bussey, G., Marconi Co., Coventry UK 2000

2 *Wireless Telegraphy*, Eastwood, E. (ed.) Wiley (Halsted Press) New York 1974, has Marconi's Royal Institution lectures of June 1902 & March 1908.

3 *Scientists' reactions to Marconi's transatlantic radio experiment*, Ratcliffe, J.A., Proc. IEE, Vol.121, No. 9, Sept. 1974

4 *Wideband Signals from Poldhu?* MacKeand, J.C.B. and Cross, M.A., IEE Conference Publication 411, IEE London, 1995

5 *The Poldhu Story - Fact or Fiction*, Garratt, G.R.M, Short Wave Magazine, Vol. XXXV, October 1977 p475-476 and November 1977 p552

6 Private communications, D. Thackeray, 1996-8. A simulation shunting series inductance with stray capacitances shows similar LF/HF jigger performance.

7 *The First High-Power Transmitter at Poldhu*, Thackeray, D., AWA Review, Vol.7, 1992 p29-45

8 *Experimental Radio Engineering*, Morecroft, J.H., Wiley, New York 1931 has a short comprehensive treatment of spark transmitters, but his classic text is *Principles of Radio Communication*, Wiley, NY. Deliberate multi-frequency spark operation was first described by J.A. Fleming (The Times, London 4 Oct 1900). A recent article on spark systems is *Arcs and Sparks*, Pt.1, Byron, W.J., Communications Quarterly, Vol.4, Nr.2, Spring 1994

9 Aiya [see bibliography]
10 Skomal
11 *The Principles of Electric Wave Telegraphy and Telephony*, Fleming, J.A., Longmans Green, London, 1910
12 *Wireless Telegraphy*, Collins, A.F., McGraw, New York, 1905
13 ELNEC
14 *History of Radio Telegraphy and Telephony*, Blake, G.G., Chapman & Hall, London 1928. *Letters to the Editor*, Electrician, V.51, 1903, Guarini, E., p360, Solari, L. p502 There was much dispute about the invention of the mercury-iron coherer, although not it appears, among the possible inventors. Solari remained a firm friend of Marconi into their later years.
15 *Early Radio Wave Detectors*, Philips, V.J., Peter Peregrinus, London 1978
16 *On Coherers*, Eccles, W.H., Electrician, V.65, p724-727 & 772-773, 1910 reports 10 nW sensitivity specifically for a mercury/iron coherer.
17 *Electrons, Electric Waves and Wireless Telephony*, Fleming, J.A., The Wireless Press, London 1923
18 CCIR Report 322
19 *Some Measurements on the Directional Distribution of Static*, Harper A.E., Proc. IRE, Vol.17, Nr.7, p1214-1224, July 1929, also Kessler, p28, and Watt, p468
20 *Polarization and Fading Near the LUF*, Hayden, E.C., see Jones
21 Watt, p332 and *On the Influence of Solar Activity on Radio Transmission*, Austin, L.W. & Wymore, I.J., Proc. IRE, Vol.16, Nr.2, p166-173, Feb. 1928, also Davies (NBS Monograph 80)
22 *Radio Noise from Thunderstorms*, Horner, F., see Saxton
23 *The Lightning Current*, Ogawa, T., see Volland, p39
24 *Low Frequency Noise*, Volland H., see Volland, p202
25 A classic treatment is *Auditory Patterns*, Fletcher, H., Reviews of Modern Physics, V.12, p47-65 Jan. 1940. See also *The Masking of Clicks by Pure Tones and Bands of Noise*, Hirsh, I.J., Rosenblith, W.A., and Ward, W.D., J. Acoustical Society of America, V.23, No.5, Sept. 1950 p631-637. Aural "click" spectra appear in *Bases of Hearing Science*, Durrant, J.D. and Lovring, J.H., Williams & Wilkins, Baltimore 1995 p97

CHAPTER XI

LAST SIGNALS FROM KHAQQ

Amelia Earhart's last flight, from Lae in New Guinea, had as its immediate destination Howland Island, a mid-Pacific speck only a few miles long and with no significant elevations, and an essential Pacific landing ground on her round-the-world flight. It would have been a mark of excellent navigation by Fred Noonan, Earhart's navigator, to pinpoint Howland without modern aids, and although their attempt failed, it will become clear that it failed only by a very narrow margin. [Fig. 11.1] The importance of radio communication and radio navigation to this leg of her route was recognized, and the Electra, using call KHAQQ, was provided with then modern radio equipment. U.S. Coast Guard Cutter *Itasca* was located off Howland, fitted out with up-to-date equipment, and awaiting her arrival with a well manned radio watch. There is still some doubt whether Earhart heard *Itasca*, but the one-way traffic from KHAQQ to *Itasca* was quite thoroughly recorded. [Fig. 11.3] Her radio-heralded arrival did not however occur; there has since 1937 been constant speculation on the fate of her Lockheed Electra NR16020; the crash location is as yet unidentified. It was suggested to the author by Maj. David Dunlap, a military pilot, that the likely crash site might be better located by using the considerable amount of radio engineering and meteorological data available, most of it relating to the radio equipment, [1] but some also to onsite weather and ionospheric conditions.

Many previous authors have examined the radio data, and several [2] have looked at the records from the stand-point of their long experience in aviation and marine radio. On the other hand, many appear, even today, to feel that short-wave radio transmission is maybe no more than the luck of the draw, statements on "freak" propagation abounding. In the years following the Electra's disappearance, this was fair comment, but in the last decade or so with the advent of the personal computer, calculations on the transmission of radio waves have become very much more accessible. Today therefore we can make quite firm determinations which, as will be seen, very much delimit the area where the flight ended, and describe it with some fair degree of probability, eliminating many of the speculative possibilities previously offered.

THE PROBLEM

While there is not much doubt as to the equipment used by KHAQQ and *Itasca*, many more questions had to be answered in establishing the parameters affecting the recorded radio transmissions from the Electra. Most of these have been amenable to calculation but some remain in the area of judgement, and in these latter instances we have found no real disagreement with the authors mentioned above. There is quite a list of possible variables to consider and first we will list them and then comment in detail on each one.

Locations and times.

Times have been well researched by others and we have used those results. [3] The *Itasca*'s position is well established, and analysis of the earlier transmissions from KHAQQ is consistent with navigational assumptions which we and others have made about the Earhart/Noonan flight path. In the case of the later transmissions, of course, location is what we are trying to establish. There is also, for the Electra, the matter of altitude. This can be important for the later, more local, positions, and we have been to some trouble to attempt to verify probable aircraft height at the presumed time of closest approach to Howland Island.

Aircraft Transmitter.

There seems to be no doubt that the frequency on which *Itasca* actually heard KHAQQ's signals was 3.105 MHz, although there has been considerable argument about the use or non-use of other frequencies. The transmitter power output has required careful analysis, as it was not only determined by the 50 watt design, [4] but would also have been extremely dependent on the characteristics of

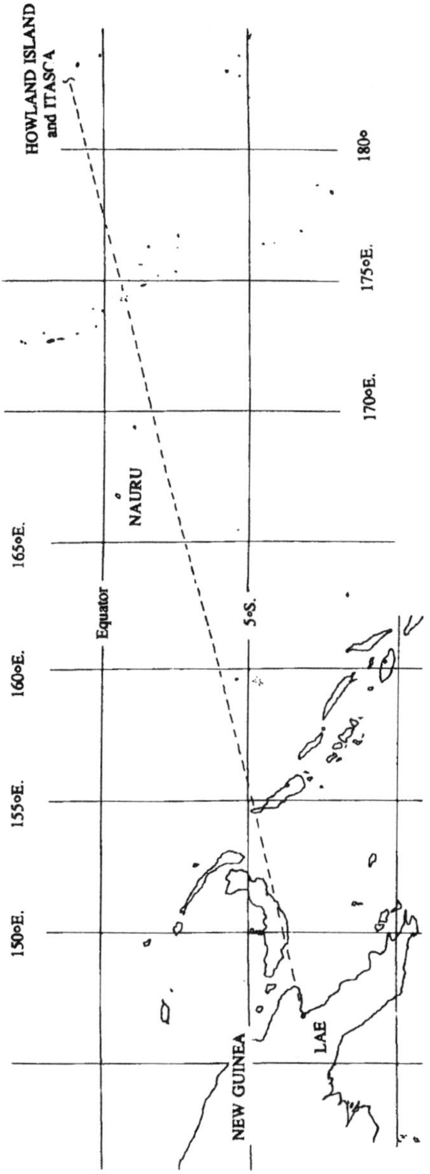

Figure 11.1 Route Map from Lae, New Guinea to Howland Island and USCGC "Itasca"

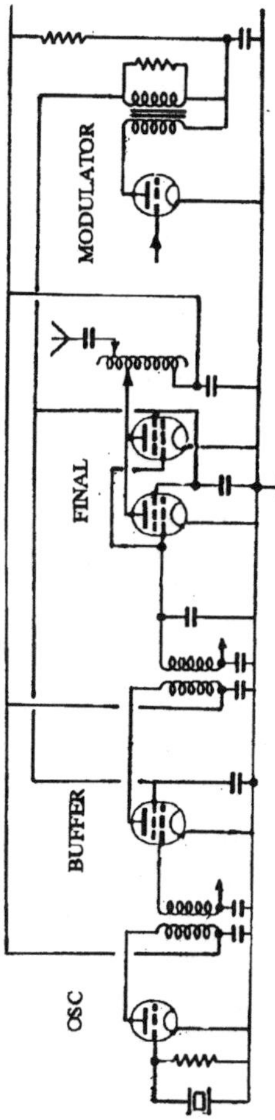

Figure 11.2 Outline Circuit Diagram of Western Electric Type 13A Transmitter

ENTRIES	TIMF
REPEATED WX ON FONE//	030
NOTHING HEARD FROM EARHART//	15
SENT WX AT	30
A A A A A A DE NRUI / HOWLAND	34/
REPEAT WEATHER ON FONE 3105//	35
EARHART HEARD FONE/ WILL LISSEN ON HOUR AND HALF ON 3105-SEZ SHE	45/4--
BROADCAST WEATHER FONE 3105//	0400
REPEATED WEA ON KEY/ 3105 KCS/	02
EARHART UNHEARD//	15/18
BRDSCT WX ON KEY AND FONE//	30/35
SENT WEATHER /CODE/FONE/ 3105 KCS--(HEARD EARHART -(PART-CLDY)----	53
-- - - - - - - - - - - - - - - - - - -	0500
EARHART UNHEARD//	15/18
SENT WX AT- FONE KEY- 3105	30/35
AAAAAAAAAANRUI TO 7500 KCS//	40
NO HEAR DURING ((SETTING ON 3105 ALL TIMES//	45/50
SENT WEATHER / CODE AND KEY 3105 KCS/	0600-05
WANTS BEARING ON 3105 KCS// ON HOUR// WILL WHISTLE IN MIC	14//
ABOUT TWO HUNDRED MILES OUT// APPX// WHISTLING// NW	15
AAAAAAAANRUI (7500)-(LISTENING THRU ON 3105)	30-
AAAAA NRUI --	35
CALLED EARHART ON 3105 VOICE//	36
AAAAAAAAAA NRUI NRUI TO //PSE ACKN 3105	41
SE TAKE BEARING ON US AND REPORT IN HALF HOUR--	45
I WILL MAKE MOISE IN MIC -ABT 100 MILES OUT	46
AAAAAAAAAAA 7500	0705/
AAAAAAAAA - 3105	08/12
AAAAA NRUI ETC// 7500	12-14
AAAAAAA NRUI 7500 3105	14-16
FONE TO EARHART/ CANNOT TAKE BEARING ON 3105 VY GOOD/ PLESE SEND ON	
500 OR DU WISH TAKE BEARING ON US / GA PSE// NO ANSWER	18
KHAQQ DE NRUI A-S GA 3105 / UNANSWD	19-24
KHAQQ FM ITASCA PLS GA ON 3105 KCS / UNANSWD / CRM/DC NOW ON THE	25
KHAQQ DE NRUI A-S GA 3105 / UNANSWD / NRUI'S MAIN D/F 500 KCS	26-9
KHAQQ FM ITASCA PLS Y OUR SIGS ON KEY PLS / UNANSWD	30
KHAQQ DE NRUI A-S	31-4
KHAQQ DE NRUI A-S, 7500 KCS	35-40
KHAQQ DE NRUI A-S, 3105 KCS	41
KHAQQ CLNG ITASCA WE MUST ON XX YOU BUT CANNOT SEE U BUT GAS IS	
RUNNING LOW BEEN UNABLE TO REACH YOU BY RADIO WE ARE FLYING AT A	
1000 FEET	42
KHAQQ DE NRUI R MSG R QSA 5 R A-S (500 ES 3105) GA / UNANSWD	43-6
KHAQQ DE NRUI R MSG R QSA5 R AS-S (3105)	47-8
KHAQQ FM ITASCA UR MSG OK PLS Y WID A3, A3/3105 / A-S FM NRUI A2	49-50-57
KHAQQ CLNG ITASCA WE ARE CIRCLING BUT CANNOT HR U GA ON 7500 WID A	
LNG COUNT EITHER NW OR ON THE SKD TIME ON ½ HOUR (KHAQQ S5 A3)	58
KHAQQ DE NRUI A-S, 7500 / GA 3105	
KHAQQ CLNG ITASCA WE RECD XX UR SIGS BUT UNABLE TO GET A MINIMUM PSE	
TAKE BEARING ON US AND ANS 3105 WID VOICE / NRUI DE KHAQQ LNG DASHES	
ON 3105 / NRUI2 DE NRUI P AR	0800-3
NRUI2 GIVING TO DPE TT NO SIGS ON 3105 ES IMPOSSIBLE TO WRK ITA / R	04
KHAQQ FM ITASCA UR SIGS RECD OK WE ARE UNABLE TO HEAR U TO TAKE A	
BEARING IT IMPRACTICAL TO TAKE A BEARING ON 3105 UR VOICE HW DO U	
GET TT GA / UNANWD, 3105	05
KHAQQ DE NRUI GA ON 3105 OR 500 K, 7500	06

Figure 11.3 Original of Itasca's Radio Log

the antenna in use. Output matching and modulation systems used in this Western Electric equipment, both quite unusual when compared with more recent radio engineering practice, also affect the calculations and were therefore reviewed.

Aircraft antenna.
Four antennas were fitted to the Electra. The antenna used for transmission of short-wave signals was undoubtedly a modified vee wire.[5] Also installed were a direction finding (DF) loop (Bendix), a sense antenna under the fuselage, also for use with DF equipment, and a tail mounted trailing long-wire, apparently removed before the final flight. A computer model of plane and antenna was used to analyse the vee wire, and contributed to the transmitter power output calculations and also, separately, provided the actual antenna gain, the extent to which it can concentrate the supplied power into directions desirable for communication.

Propagation.
Distant signals, the earliest logged notations of KHAQQ's approach to *Itasca*, were by sky-wave, and the program *SNAPmax*[6] was used to show that these log entries were consistent with what we today know of longer distance short-wave propagation in the Howland area at the time and date in question. Local signals, as represented by the last *Itasca* log entries, were clearly ground wave and have been analysed in depth and detail. Sky wave propagation was rapidly disappearing as the Electra approached Howland, soon after local sunrise.

Ship's antenna.
Itasca, a typical Coast Guard cutter, had quite a panoply of wires and stays, and among them at least two antennas. Use of the traditional ship's Marconi vertical antenna has been assumed, although it rapidly became evident that this was a very minor factor in the calculations.[7]

Ship's receiver.
The receiver used on board the *Itasca* was the Coast Guard CGR-32-1,[8] and we were fortunate enough to locate an example of this equipment,[9, 24] maybe slightly better known as the RCA AR-60, complete with its original operating and service manuals and circuit diagrams.

Received signal strength.
"Signal Strength 5" (S-5) reported by a human operator, and based on audio level in headphones or a loudspeaker, is not readily translated to a received voltage. That depends not only on the strength of the received signal, but also on the receiving antenna and its connections, the receiver, its state of adjustment, the received frequency, and on the "ear" of the operator. The definition of S-5 is therefore based on the perception of signal strength relative to noise level and almost independent of the characteristics of the receiving equipment.

These are the many variables which we tried to pin down, with varying success, but all contribute to the final proposed results. As in any work of estimation, there is some advantage to be gained by breaking the analysis down into its smallest components, so that the likely positive errors in one may negate likely negative errors in another. The magnitude of the uncertainty is here important and is also addressed.

ANALYSIS OF THE COMPONENTS

Performance of the Aircraft Transmitter.
The transmitter fitted in Amelia Earhart's aircraft was a modified Western Electric Type 13A, [Fig. 11.2] advertised as, and without doubt capable of, delivering 50 watts of radio frequency power into an appropriate load. [10] We located circuit diagrams and operating and maintenance instructions for this equipment [11] and also discussed the system with a retired Western Electric engineer who had some involvement with the original installation, [12] The plate tank and antenna coupling circuit was similar to many small transmitters of the 1930s. Antenna and plates connected to movable taps on the tank coil, and except for low frequencies (500 KHz) there was no series loading coil for a "conjugate match" such as one might expect to find today. The vee wire R.F.loading proved to be extremely light. It was therefore necessary to calculate radiated power from first principles, [15] rather than accept Western Electric's data for standard installations. They noted that the resistive part of antenna impedance had to be greater than 4 ohms in order to obtain 50 watts R.F. output.

The final amplifier comprised two recently designed Western Electric 282A pentode tubes, operated in parallel, and supplied from a dynamotor at a nominal 1050 volts D.C. to the plates, say 1100 volts unloaded. Thus the maximum possible unloaded voltage swing in this amplifier, considered as a Class C operation, would be 2200 volts A.C. peak-to-peak R.F., or very close to 740 volts R.M.S. However, in R/T mode (double-sideband or A3 radio-telephony), the 13A family had a unique method of modulation. The carrier was produced by a Western Electric 205D triode tube operating in a crystal oscillator circuit at half the output frequency, doubling in the plate tank, and driving a 282A pentode buffer at the final frequency. This buffer was transformer modulated on the screen grid, to less than 100% modulation. The 282A finals, probably Class B, were also partially screen modulated to bring the final modulation level to 100%. Audio distortion was kept "below 10%", [13] and given the use of a carbon microphone, this was then very acceptable practice. However, screen modulation is efficiency modulation and the drive and supply voltages must be set so that full radio frequency voltage in the tank circuit is achieved only on voice peaks in order to avoid overmodulation or "flat topping", although audio quality reports for KHAQQ suggest that some limiting may in fact have occurred. Allowing for modulation, the final amplifier output voltage is therefore reduced to 370 v. R.M.S. or 50% of the calculated Class C value to represent average D.S.B. A.M. telephony carrier strength.

There is some uncertainty in antenna wire size, both #12 phosphor-bronze and 0.040 in. Copperweld being likely candidates. [14] This analysis assumes Copperweld, with an impedance of 0.717 - j376.3 ohms. The coil taps provide an impedance match from the 282A plates (taken as 1500 ohms, similar to two 4X150A tubes) to the antenna, [15] and the best match, [16] assuming an unloaded tank coil Q of 250, produces a voltage step-up of 3.26. The antenna therefore sees 1206 volts R.M.S. and a current of 3.205 amps. Unfortunately, this is almost entirely reactive charging current, derived from a virtually constant voltage source, and could produce in the 0.717 ohm resistive component a real antenna power of only 7.36 watts, a result verified by equivalent circuit analysis taking into account the tank coil losses. [17] The radiation resistance is even lower, about 0.156 ohm, the very significant antenna efficiency (copper) loss being included in calculated antenna gain. At other frequencies, and with other antennas, the Western Electric 13A manual notes maximum and minimum antenna currents of 1.0 to 6.0 amps. A short antenna at a low frequency must have a high capacitive reactance and a low radiation resistance, so there is no surprise in these results.

Gain of the Aircraft Antenna.

A method-of-moments analysis program was applied to the complete antenna, [18] namely the vee-wires and the metal skin of the aircraft together. It was no surprise to find that at a frequency of 3.105 MHz, where the wavelength is approximately 315 feet, the antenna behaves as an electrically short wire. Antennas are electrically short if their maximum effective dimension is much less than half a wavelength, and indeed it was. Low gain results, and this system is no exception as the simulation results demonstrated. [Fig. 11.4] A half wave dipole, which is no sophisticated antenna, has a gain of 2.15 dBi, that is, it is 2.15 dB better in its best direction than the standard theoretical reference or isotropic antenna. The Electra's vee wire system at 3.105 MHz had, in comparison, a gain of 0.245, or 6.11 dB worse than an isotropic reference for dead ahead ground-wave transmissions, and was only slightly better for sky-wave operations. This is no reflection on the installation, but rather an unavoidable fact of life, that on a small airplane, it is difficult to install a large antenna. The only practicable higher efficiency antenna at 3.105 MHz would have been the trailing wire, which for one reason or another was not deployed.

These calculated gains can be applied to the overall system; total field for the sky-wave cases, and vertically polarized results for the ground-wave cases. ELNEC gives a clear picture with its far field calculation, and for the Copperweld antenna we obtain a field strength of 85.7 mV/m or 98.66 dBμV/m at 1 Km, for 1 Kw in the antenna. But the power in the antenna is a mere 7.364 watts, -21.3 dB relative to 1 Kw, and so the 1 Km. field strength is only 77.36 dbμV/m. However, this assumes an all-copper antenna and airplane! Reworking the calculations for a zero resistance system, antenna power with the same voltage and current in the antenna would be 1.605 watts, -27.95 dB relative to 1 Kw, and the far field at 1 Km and 1 Kw would be 183.6 mV/m or 105.3 dbμV/m, with an actual field strength of 77.33 dBμV/m. As expected, although the antenna material makes a great difference to the antenna power, in a highly reactive small antenna system such as this it makes very little difference in radiated power.

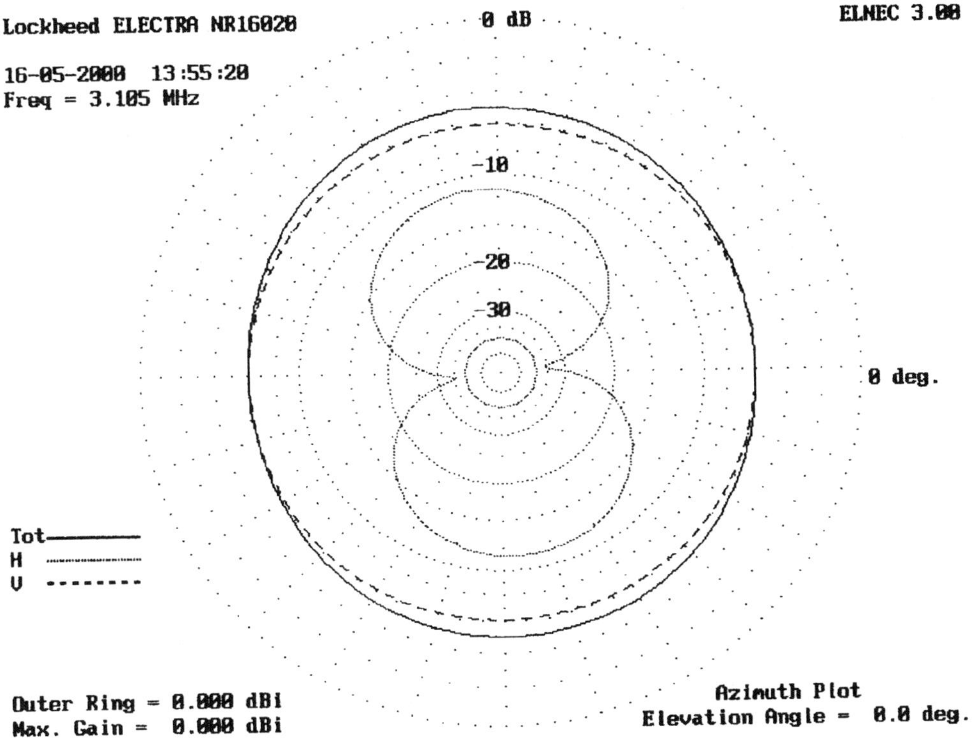

Figure 11.4 Gain Patterns - Lockheed ELECTRA Vee-wire Antenna

Propagation Analysis

The earliest reception of Earhart's signals in *Itasca*'s radio room was over sky-wave paths and was analyzed using *SNAPmax*. [Fig. 11.5] The marginal results at 1400 - 1500 GMT back up *Itasca*'s log, confirming some assumptions affecting both ground wave and sky wave modes, although prediction at this frequency, where MF shades into HF, is not as reliable as at higher frequencies. While *Itasca*'s relatively high power transmitters operating into full size antennas may well have been heard all over the Pacific Ocean, the minute radiated power from KHAQQ was certainly not, especially at her arrival time as night-time sky-wave propagation was quickly waning. We narrow down the Electra crash location on the basis of the later transmissions, reported by *Itasca* as "S-5" signal strength.

```
*---------------------------------------------------------------------------*
              SIGNAL TO NOISE (S-Units)          - SNAPmax -   by WA3ZKZ

   3 Jul 1937    10.7 cm Solar Flux   157     A Index  12       Bandwidth  6000
   Local Site   Itasca                         Latitude  1.5    Longitude  176.5
   Remote Site earh4                           Latitude  1.6    Longitude -177
   Local Beam Heading 271 degrees      Remote Beam Heading 91 degrees
                      Great Circle Distance    722  Km.
*---------------------------------------------------------------------------*

  GMT                                                                 MHz
  TIME  0    5    10   15   20   25   30   35   40   45   50   LUF     MUF
  ----  !----!----!----!----!----!----!----!----!----!----!   -----------
   0    |         ..00000*--                                     9     15.2
   1    |         ..00000**-                                     9     15.5
   2    |        ..000000*--                                     8     15.3
   3    |       ..0000011**-                                     7     15.7
   4    |      .0000111111*--                                    6     16.0
   5    |     00111111111**-                                     5     15.7
   6    |   0111111111111**-                                     3     14.6
   7    |  012222221111*-                                        2     13.3
   8    |  011222221111*--                                       2     13.6
   9    |  011222221111*--                                       2     13.4
  10    |  01122222221*--                                        2     12.7
  11    |  01122222222*--                                        2     12.7
  12    |  01122222222*-                                         2     12.4
  13    |  0112222222*-                                          2     11.2
  14    |  011222222*-                                           2     10.4
  15    |  01122222*                                             2      9.1
  16    |  012222-                                               2      7.0
  17    |  01222*                                                2      6.5
  18    |   111222-                                              3      8.0
  19    |    .0011111*-                                          4     11.4
  20    |      .00011111*--                                      6     14.3
  21    |       ..0000111**-                                     7     15.7
  22    |        ..0000001*--                                    8     16.2
  23    |         ..00000**-                                     9     15.5
  ----  !----!----!----!----!----!----!----!----!----!----!   -----------
        0    5    10   15   20   25   30   35   40   45   50   FREQUENCY
```

Figure 11.5 Sky-Wave Signal-to-Noise Prediction KHAQQ - Itasca

The simplest ground-wave case is direct line-of-sight transmission, and initially it is assumed that *Itasca* was located within KHAQQ's horizon distance. This radio horizon is somewhat farther than the geometric horizon, due to the atmosphere's refractive index gradient and it has been calculated for an aircraft altitude of 300 meters and a standard atmosphere. Although a height of 1000 ft (304.8 m) was given by Earhart in one of her transmissions, this has caused some argument. Other choices would have appealed to many pilots seeking a small oceanic island, but local weather conditions, verified by photographs taken at, or close to, the expected time of arrival on Howland Island, and meteorology based on *Itasca*'s ship's log, support her choice of an altitude beneath developing storm clouds. The radio horizon distance [19] for an aircraft altitude of 300 m and a ship's antenna height of 5 m, (from ocean surface to antenna base), was 80.8 Km. In this quasi-free-space environment, [20] the relevant criteria are the antenna losses or gains, the power in the transmitting antenna, the signal expected at the receiver site, and above all, the distance. Free-space field strength from a transmitting isotropic radiator is given by [21] :-

$$E_0 = 173\ (W_t)^{0.5}/d \quad \text{millivolts/meter (mV/m)}$$

where W_t is radiated power in kilowatts (kW), d is distance in kilometers (Km) and 173 is the isotropic constant, but an even simpler relation enables a straightforward extrapolation from the 1 Km. field strength provided by the antenna model program.

$$E_0 = E_1.d_1/d_0 \quad \text{millivolts/meter (mV/m)}$$

where E_1 is the field strength at 1 Km, d_1 = 1 Km and d_0 is aircraft distance in kilometers. The value of E_1 we have calculated as 77.36 dBμV/m or 7.38 mV/m, and therefore the field at 80.8 Km is 91.3 μV/m or 39.2 dBμV/m. This result can also be checked against published CCIR curves, [22] although they assume that both stations are at sea level. At a range of 80 Km and the standard radiated power of 1 KW, the signal strength is given as approximately 67 dBμV/m. Allowing for the fact that the actual radiated power is only 1.605 W, some 28 dB down on the standard, the receiving antenna signal would be 39 dBμV/m, 89 microvolts/meter, elevations below a critical or boundary height of about 1000 meters for a 100 m wavelength having little effect on received signal strength.[23] Tropospheric effects, although hardly calculable, were unlikely to be significant at the frequencies used.

Effect of the Ship's Antenna

A Marconi or vertical antenna was standard shipboard equipment, but *Itasca* appears also to have had a "doublet". Photographs of the ship's rigging[7] may show a doublet [not necessarily a half-wave dipole] from the mast to the foc'sle area. Such antennas, partially horizontally polarized, are less effective in ground-wave than Marconi bottom-fed verticals, with or without flat-top. Neither Coast Guard photographs nor correspondence give a positive description of the antenna actually used, but a vertical seems most likely, as its superiority over horizontal polarization for ground waves was already well-known. Antenna height, for radio horizon calculations, has been taken as 5 meters a.s.l. at antenna base level. Since the relation of reported QSA S-units to receiver radio frequency input level is very dubious at best, no attempt was made to assess gain of the ship's antenna system.

Response of the Ship's Receiver

Itasca's Coast Guard CGR-32-1 receiver handled incoming transmissions from the Electra. Coast Guard Archives eventually revealed that this high-performance radio was an RCA AR-60, manufactured in Camden N.J. by Radio Corporation of America. At least two have survived, from coast radio WCC in Chatham, Mass., and we were able to inspect one and study its specifications in detail.[24] The AR-60 was built in small numbers, probably no more than 250, originally for the top end of the amateur radio market, but most were purchased by the U.S. Coast Guard. It was a superhet, very advanced for its time (1935), of the solidest "battleship" construction, and very expensive! No sensitivity figures are quoted in the literature, but we saw that it was equipped with two pentode (6D6) R.F. amplifiers, one of which was out of circuit in the lower bands covering our frequency of interest. Also, 3.105 MHz was almost mid-range in the tuning of the relevant band, and it is clear that at this fairly low frequency, noise figure was no problem, and neither was sensitivity at the signal levels we are considering.

Receiver bandwidth was estimated from RCA's curves as 6 KHz. The *Itasca*'s Chief Radioman observed, in later comments, [25] that a more modern receiver might have enhanced results. While this could be valid for the earlier sky wave signals, it would have no bearing on the large signal ground wave reception on which the final distance calculations are based.

ANALYSIS OF THE OVERALL SYSTEM

There is no satisfactory way to determine a numerical meaning in terms of received voltage for "S-5", the best signal strength reported by *Itasca* on any of the KHAQQ transmissions, but a contemporary British Naval document describes 100 microvolts/meter at H/F (high frequency) as "quite a strong telephony signal" and 10 - 20 microvolts/meter as providing "satisfactory reception". [26] This data point does however provide one very useful point of reference for the following noise level calculations.

Radio-Frequency Noise Level Basis

A different approach to the problem can get around the difficulty. The R.C.A. AR-60 receiver had no "S-Meter" for visual indication of carrier strength, and so the S-5 assessment was a purely subjective judgement by *Itasca*'s experienced radiomen. The "ear" of the operator will identify an S-5 signal, not by its absolute signal strength, but by its margin over the ambient noise level. In other words, a signal reported as S-5 in high ambient noise conditions would have to be stronger compared with the received signal voltage than an S-5 signal reported in a low noise situation. And, if the ambient noise field strength, from atmospheric, man-made and galactic sources, is known, it can be compared directly with the field strength of the received signals. The receiver's sensitivity, not to mention its state of alignment, coupling losses from ship's antenna to receiver, and the ship's antenna itself, with any possible transmission line, being identical for both noise and signal, now all cancel out of the equation.

Howland Island Atmospheric Noise Level
Ambient atmospheric noise levels for the place, and date of KHAQQ's final transmissions, were obtained from CCIR Report 322, [27] and the probable local noise level, F_a, at 1 MHz for the appropriate local time block (Summer 0400 - 0800) was 60 $dBkT_0b$, falling to 16 dB for the time block 0800 - 1200. *Itasca* recorded local sunrise at 1745 GMT, and the last local transmission occurred at 0740 local time, 1928 GMT, 0758 at Lae, New Guinea, so the noise values must be interpolated to get Howland Island noise level at the actual KHAQQ transmission times. Two major noise sources are evident, one in the South China Sea and the other in the Coral Sea. [27, 28, 29]

The main variations take place around sunrise, [27, 29] and if noise is mainly ionospherically propagated, transition timing depends on solar illumination at path control points. If the source is East of the location then transition occurs before local sunrise, as D layer absorption increases between location and source; if sources are North or South, transition is set by local sunrise; if to the West, it occurs later than the location sunrise. While local noise and other remote sources contribute, the China Sea source overshadows them, and we need to examine the propagation from there to Howland. A substantial fall-off can be attributed to that failing propagation, reducing the level to a constant 38.6 dB at about 1945 GMT. A URSI curve [29] shows that around local sunrise, storm noise also decreases at the source at some 15 dB/hr, further reducing level after 2000 GMT, but atmospheric noise is already low enough to be insignificant compared with other sources.

With a sunspot number of 113, the F-layer critical frequency was around 6 MHz, and galactic noise was therefore absent. The receiver front end noise level at this relatively low frequency was very unlikely to be worse than 10 - 15 $dBkT_0b$, [24] and both these sources are therefore neglected.

***Itasca* Man-Made Noise Level**
For the man-made contribution, we have to estimate the likely radio noise level from local electrical systems on a steam vessel of inter-war-years design. What is known is that the *Itasca*, a member of the *Lake* class commissioned in July

1930, [30] had an innovative 3000 s.h.p. A.C. turbo-electric drive. Also, while Earhart was making her approach *Itasca* was station-keeping at low propellor r.p.m., [31] and therefore the exciters for the synchronous motor drive and for the main alternators would be active noise producers. However, with good installation practices and minimal electrical equipment in today's terms, and the very considerable shielding effects of machine casings and conduit, and bulkheads and deck plating of a steel ship, cited at 70 dB, [32] it is considered that ship's noise could be as low as the range given by CCIR for rural areas. Topside-generated interference ("rusty bolt" rectification of off-frequency signals from other shipboard transmitters) could have been considerably higher, [33] but being under the direct control of the Chief Radioman, it could also have been mitigated, and most probably eliminated. While we did not find any records of actual noise levels on these Coast Guard cutters, one retired Coast Guard radioman has said that they certainly did not have a reputation as noisy ships.

Although there are few records of typical ship's noise level, we did find three independent sources. We already know [26] that on a British warship of the 1930s 100 μV/m was considered to be a signal adequate to produce "quite a strong telephony signal". Equating this to S5, and an SNR of 35 dB, we find that for a 6 KHz bandwidth at 3.105 MHz, F_t is 52.9 $dBkT_0b$. The U.S. Naval Electronics Laboratory, in a 1970 report, [33] noted a "quasi-minimum" shipboard noise level at various frequencies. Interpolating at 3.1 MHz we have 46.5 $dBkT_0b$, and taking this measurement as a lower decile, we find a correction [34] of 8.8 dB to median level for man-made noise on this frequency. The resulting F_t is 55.3 $dBkT_0b$. Another paper cited "coupling factors" for a variety of merchant ships of the 1930s and early 1940s, attempting to show the attenuation between noise sources and induced signals in the ship's antennas, but unfortunately only one set of usable numbers for source strength and distribution was given. This was for a series of measurements of noise field strength at a number of frequencies and deck locations on board the famous Cunard ocean liner *R.M.S. Mauretania.* [32] From this listing, selecting a 3 μV/m value in 10 KHz bandwidth as a best fit for probable antenna location, and converting to 6 KHz bandwidth at 3.105 MHz, we obtain an F_t of 55.11 $dBkT_0b$. The three values are remarkably close, and taking a simple average, a suggested ship's noise level is 54.4 $dBkT_0b$.

Total Noise and Required Signal Strength

Considering the 1928 GMT signals as an example, atmospheric 1 MHz noise F_t would be 54 dB, or 45 dB at 3.1 MHz.[27] Assessing commercial circuits, where circuit availability is relevant, upper decile noise may be substituted for mean noise, but for our purposes it seems more appropriate to use mean level, omitting this D_u correction. Ship's noise therefore controls at 54.4 $dBkT_0b$, for a total of 54.9 $dBkT_0b$. The noise level is specified in a short vertical antenna, translating to R.M.S. field strength at the frequency and bandwidth in use by the equation:-

$$E_n = F_t - 95.5 + 20 \log_{10} f + 10 \log_{10} b = F_t - 47.9 \text{ dB}\mu\text{V/m}$$

where f is the received frequency in MHz and b is the received noise bandwidth in Hz. The most likely received noise level E_n is therefore 7.0 dBμV/m. Double-sideband A.M. (A3) is not now in widespread commercial use, but earlier editions of CCIR published data[36] specified the signal-to-noise ratios [SNR] required for "marginal quality", "D.S.B. commercial" and "Broadcasting" signals, as 27, 35, and 47 dB, D.S.B. commercial grade most probably decribing S-5 in U.S. Coast Guard practice. A lower SNR might have been reported S-5 by radio officers who had spent the night straining for a weak sky-wave signal, but *Itasca* handled other radio traffic during most of the period, and normal standards probably applied. For an A.M. transmission in a 6 KHz bandwidth, the cited SNR of 35 dB would require a signal field strength of 42.0 dBμV/m.

Aircraft Distance

This completes the basic information needed to determine the free-space distance between the Electra and the *Itasca*, and it can be calculated from:-

$$d_0 = d_1 \,.\, 10^{(E1 - Es)/20} \text{ Km} = 10^{(77.36 - Es)/20} \text{ Km},$$

where d_1 is a reference distance (1 Km), E_1 is the far field strength at distance d_1 in dBμV/m, and E_s dBμV/m is required signal strength at the receiving location. This calculated distance must then be adjusted as in CCIR Recommendation 368 for oceanic ground-wave. The CCIR curves[35, 37] are for stations at ground level, but are applicable below critical heights of about 1000 meters. The distance may be calculated from the curve-fit equation below,

$$d_c = d_0 - 0.0427\ (d_0)^{1.4168}\ \text{Km.}$$

where d_c is a subtractive correction for a wavelength of 100 m., to be applied to the free-space value d_0, both distances in Km. At 1928 GMT the most probable free-space distance was 58.6 Km, corrected to 44.9 Km, with high and low limits of 76.2 Km and 25.8 Km for one standard deviation. Other results are shown in Table I. Similar values can be derived for signals at 1815, 1912 and 2014 GMT.

Deviation	-2 s.d.	- 1 s.d.	Mean	+1 s.d.	+2 s.d.
Noise F_t $dBkT_0b$	44.1	49.5	54.9	60.3	65.7
Noise E_n dBμV/m	-3.8	1.6	7.0	12.4	17.8
Signal dBμV/m at 35 dB SNR	31.2	36.6	42.0	47.4	52.8
Free-Space Range Km	203.2	109.1	58.6	31.5	16.9
Ground-Wave Range Km	123.7	76.2	44.9	25.8	14.5
Probability %	95.0 →	68.3 →	Most probable	68.3 ←	95.0 ←

Table 11.1 Signal strength, range and probability at 1928 GMT.

At 1815, KHAQQ's signal report was "fair reception", taken as the CCIR 27 dB "marginal" SNR. The 1 MHz noise was 60 $dBkT_0b$ (F_t 51 dB at 3.1 MHz), combining with ship's noise to a total of 56.2 $dBkT_0b$ or 8.4 $db\mu V/m$. Signal level therefore would be 35.4 $dB\mu V/m$ for a most probable range of 85.0 Km with higher and lower limits at 1 s.d. of 136.3 Km and 50.6 Km. The 1 MHz noise level at 1912, most likely her time of closest approach to Howland Island, was 57 $dBkT_0b$ (F_t 48 dB at 3.1 MHz) for a combined total noise of 55.3 $dbkT_0b$ or 7.5 $dB\mu V/m$. As the signal was also regarded as S-5, at 35 dB SNR this required 42.5 $dB\mu V/m$ signal strength, with a most probable range of 42.6 Km and limits of 72.5 Km and 24.9 Km at 1 standard deviation.

At 2014 GMT, F_a is estimated at 35 $dBkT_0b$, (F_t 22dB at 3.1) MHz, which is negligible compared with ship's noise at 54.4 $dBkT_0b$ or 6.6 $dB\mu V/m$. Again, signal reports were S-5, for a signal strength of 41.6 $dB\mu V/m$, a most probable range of 46.7 Km, and limits of 78.9 Km and 26.8 Km at 1 standard deviation.

Examining probable sources of error in these calculations, one major known variance is that stated by CCIR 322 for atmospheric noise level. The most significant unknowns are the ship's ambient noise level, and the SNR attributed to the "S-5" signal described by the *Itasca*'s radiomen. The standard deviation for the CCIR noise values is stated as 4 dB, that for signal-noise ratio has been taken as 3 dB, with 2 dB for ship's noise level. This results in a total effective noise s.d. of 5.4 dB, allowed for in Table 11.1, which shows computed results at the mean or most probable distance, and at one and two standard deviations above and below the mean.

Conclusion

Amelia Earhart's Electra aircraft came well within radio horizon distance of Howland Island. Radio data places her most probable location during her last four transmissions to the waiting Coast Guard cutter *Itasca* at the distances in Table II. These results reinforce the often stated view that closest approach was at about 1912 GMT, and that she thereafter headed away from the island.

Distance and Time GMT	1815	1912	1928	2014
Kilometers.	85.0	42.6	44.9	46.7
Statute miles.	52.8	26.5	27.9	29.0
Nautical miles.	45.9	23.0	24.2	25.2

Table 11.2 Time and most probable distance.

Her point of closest approach, at 1912 GMT, was probably between 15 miles (13.4 n.m., 24.9 Km) and 45 miles (39.1 n.m., 72.5 Km) distant. While it is most unlikely that the ship's watch could have seen such a small airplane as the Electra as far as 10 miles (8.5 n.m., 16 Km) distant, the visible range of Howland Island, with *Itasca*'s smoke plume, from an aircraft at 1000 ft. might have been 15 miles (13.4 n.m., 24.9 Km), or even more, with fresh eyes and no haze, but the aircraft was not sighted and neither, obviously and unfortunately, was the ship or the island.

The *Itasca*'s crew believed that the Electra was almost on them, and others have come to the same conclusion based on their own experience interpreting the *Itasca*'s logs. It is very tempting to conclude from the radio evidence that in her search for Howland Island, she was running on bearing 157-337 and passed by *Itasca* only some 25 miles (22 n.m., 40 Km), or less, short of her target.

Notes and References.

1 *The Coast Guard at War - Communications - Vol. I*, Historical Section, Public Information Division, U.S. Coast Guard Headquarters, Washington DC 1947, p148-151, also refs. 3, 4a, 4b, 4c, 4d, 4e, 7, 9, 12 & 24 below.

2 *Amelia Earhart Disappearance*, Long, E.M., Amelia Earhart Symposium Record, Aviation Journal, 1983
Amelia Didn't Know Radio, Gray, A.A., Naval History - United States Naval Institute, p42-49, Nov./Dec. 1993

3 *Sound of Wings*, Lovell, M.S., St. Martin's Press NY 1989, Appendix A

4a *A Three-Frequency Radio Telephone Transmitter for Airplanes*, Tinus, W.C., Western Electric house journal c.1936 (from W.E. archives.)

4b *Aviation Communication Equipment Type 208*, Western Electric, 22 pages, n.d.

4c *Instruction Bulletin No. 650, Radio Transmitter No. 13A and No. 100 Radio Transmitter Mounting*, Western Electric 1935

4d *Aircraft Radio and Electrical Equipment*, Morgan, H.K., Pitman & Co., NY 1941

4e *The Radio Manual*, Sterling, G.E., Van Nostrand, NY 1938

5 see ref. 2

6 *SNAPmax*, Tyndar Press, POB 236, Montchanin DE

7 USCG *Itasca* antenna - photographs from Coast Guard Historian's Office.

8 see ref. 1

9 *De Luxe Amateur Set has Continuous Band Spread*, RCA Service News, Nov. 1935

10 see ref. 4b, c & d

11 see ref. 4c & d

12 telephone communication, Grosselfinger, W., W2ATQ

13 see ref. 4a

14 *Constants of Fixed Antennas on Aircraft*, Haller, G.L., Proc. I.R.E., Vol.26, No.4, Apr. 1938, p415-420

15 Terman.

16 *Radiotron Designer's Handbook*, ed. Langford-Smith, F., RCA Harrison NJ 1957, p920

17 *Radio Receiver Design*, Part 1, Sturley, K.R., Wiley, New York, 1947 p93
18 ELNEC
19 David & Voge, p95
20 see ref. 19 p54
21 see ref. 19 p3
22 Braun, citing CCIR 368; see also ref. 19 p78
23 Picquenard, p77,78; see also ref. 19 p92
24 *Instructions for Commercial Receivers (High Frequency), Types AR-60-R, etc.* IB-23297, RCA Victor Div., RCA Mfg Co., Inc., Camden NJ 1935, & see refs. 1 & 9
25 letter Bellarts, L.G., to Goerner, F., 15 Dec 1961, National Archives.
26 *Admiralty Handbook of Wireless Telegraphy, Vol. 2, Wireless Telegraphy Theory*, His Majesty's Stationery Office, London 1938, secn. R.30
27 CCIR Report 322
28 Jordan, also Terman.
29 *Radio Noise from Thunderstorms*, F. Horner, see Saxton, p175-180, and URSI Special Report No. 7.
30 *U.S. Coast Guard Cutters & Craft of World War II*, Scheina, R.L., Naval Institute Press, Annapolis MD 1983, p25-29
31 Ship's log, U.S.C.G.C. *Itasca*, National Archives.
32 *Radio Interference in Ships*, Pearce, S.F., Engineer, v185, 4807, p251-263, London, March 13 1948
33 *Naval Shipboard Communication Systems*, Kim, J.C., Muehldorf, E.I., Prentice Hall, Englewood Cliffs NJ, 1995 cites Gustafson. Very similar quasi-minimum noise values are also cited in Goodman, p346.
34 *Atmospheric Noise and its Effects on Telecommunication System Performance*, Spaulding, A.D. cites Hagn & Sailors' values for man-made noise lower decile margin D_l at a range of frequencies; see Volland.
35 see ref. 19, p78, Fig. 4.12b. (100 m. curve-fit). Also p92 Fig. 4.19
36 Compendium of the Technical Recommendations .... A more recent compilation, not including A.M. data, is CCIR Recommendation 339-5, data from which is also published in ref. 22.
37 CCIR Recommendation 368-2. Also ref. 19 p78 and ref. 22

CHAPTER XII

SOME AMATEUR RADIO CONTACTS

Finally, we'll take a look at a few amateur contacts, quite various in character, to help us see how changes in conditions can control propagation. First we will examine a route from the US East Coast to Western Australia, and see how it is affected by time, by frequency and also by the selection of short or long path. A route from the USA to Northern Europe, intersecting as it may well, the polar auroral zone, can show the effects of ionospheric storm conditions. And lastly we take a very brief look at a typical ground wave route. The simplest approach to these situations is with a computer software program, and for this purpose I have, unsurprisingly, used *SNAPmax*. While this strongly calculation-oriented program offers no sophisticated graphics, it does provide its user with an analytical approach to the question why a proposed route may or may not suffice, in the following and many other circumstances.

Eastern Seaboard U.S.A. to Western Australia

This long-haul route from a transmitter in Delaware to a receiver in Perth, passing over the North American continent, makes its way across the Pacific Ocean and the Australian sub-continent, sampling equatorial conditions on the way. Date and time are for a winter late-night session in Delaware, while our colleague in Perth is enjoying the Southern summer. Ionospheric conditions are set by a solar flux of 150 and a global A index of 15, while local man-made noise in Perth is taken as the CCIR's *residential* level. The mode is hand sent CW at 20 words per minute, receive bandwidth is 250 Hz and the operating frequency is 14.05 MHz.

Transmitter power is 100 watts, and both stations are assumed to be using three element Yagi beams, the American one on a 60 ft tower and the Australian counterpart on a 40 ft mast.

```
*---------------------------------------------------------------------------*
              CIRCUIT QUALITY          - SNAPmax -   by WA3ZKZ

 10 Mar 2000  10.7 cm Solar Flux  150     A Index  15      CW modulation
 Local Site  Test Circuit #1 W.A. (VK6)  Latitude -32      Longitude -116.5
 Remote Site Delaware                    Latitude  39      Longitude  75.5
 Local Beam Heading 51 degrees      Remote Beam Heading 302 degrees
                --- Great Circle Distance 18680  Km.  ---
*---------------------------------------------------------------------------*

 GMT                                                                 MHz
 TIME  0    5    10   15   20   25   30   35   40   45  DAY     -- MUF --
 ----  !----!----!----!----!----!----!----!----!----!   *     -----------
  0    |             .PPM**--                           **.        18.9
  1    |             .P**-                              **.        16.6
  2    |            .*--                                **.        14.4
  3    |           *-                                   **.        12.2
  4    |         *-                                     **.        10.4
  5    |         *-                                     **.        10.8
  6    |         .*-                                    **.        11.4
  7    |         .**-                                   *..        11.9
  8    |        .P**-                                   *..        11.8
  9    |       .PP*--                                   *..        11.7
 10    |        .**-                                    *..        11.0
 11    |         *-                                     ...        10.4
 12    |         *-                                     ..*        10.6
 13    |          .P*-                                  ..*        13.1
 14    |          .PPM**-                               ..*        15.8
 15    |          .PPMMGG**--                           ..*        18.6
 16    |         .PPMMGGGG**--                          ..*        19.2
 17    |         .PMMGGG*--                             ..*        17.2
 18    |        .PPMM**-                                ..*        14.8
 19    |        .PMM*-                                  .**        13.3
 20    |         .P**-                                  .**        12.8
 21    |          .PPMM**--                             .**        16.9
 22    |            .PPMMGG**--                         .**        20.2
 23    |             ..PMMMG**---                       ***        21.7
 ----  !----!----!----!----!----!----!----!----!----!  LMR    -----------
       0    5    10   15   20   25   30   35   40   45         FREQUENCY
```

Figure 12.1 Circuit Quality (Poor, Medium, Good) W3 - VK6

A preliminary view examines the circuit quality scan (Figure 12.1), and right away we can see that results are only marginally likely to be satisfactory at our chosen time, with just a slim chance of a contact. [Thirty meters would have been better, but still rated as *Poor* propagation.] To see why, we could inspect the signal strength, loss and noise bar charts, but let's look instead at the more detailed analysis of Figures 12.2. The transmitted signals will make six hops of 3113 Km each, leaving Delaware on a great circle bearing of 302° almost exactly to the North-West, for a route length at ground level of 18680 Km and a slant distance, the way the signal travels with all its ups and downs, of 19653 Km. The first ionospheric reflection is somewhere over Minneapolis, the mid-point of the path is not far from Midway Island in the Pacific Ocean, most of the terrestrial landings being in the Pacific, and the final encounter with the F layer is over the Australian Great Sandy Desert. Arrival in Perth is just about from the North-East, angles of take-off and arrival being 4.7°. We had specified a lower minimum angle than this, but if our antennas really won't support such a low angle, because of buildings, trees, hills (yes, there are a few hills in Delaware!) or other obstructions, then another path is described, this one with seven hops each of 2668 Km, take-off and arrival angles of 7.6° and intermediate reflections which are somewhat different from the first case.

If there is any transmission practicable on this route, signals are not too bad, with a signal-noise ratio of 15.7 dB, a respectable level somewhat above the mid-range SNR of 7 dB and only just a little below the 17 dB that we would call *Good*. However, our downfall will probably come when we see that the critical frequency is around 3.7 MHz for a MUF prediction of only 11.8 MHz. There is a possibility that it could be as high as 14.8 MHz, but on the other hand, there is also the possibility that it could be as low as 9.8 MHz. At our selected operating frequency of 14.05 MHz, *SNAPmax* calculates a probability of only 16.8% that we can establish a circuit. We can also see various other details, some interesting, and some useful too! The F layer heights are probably for most of us only in the *interesting* category, but the noise levels and signal losses may show how to improve our chances. Distance loss we must live with. Earth reflection losses are slight, due to the long oceanic traverse. Absorption and over-MUF losses may be worth consideration - to change these we probably need to consider another time of day, or another frequency. Finally, a plus, it looks as if our signal path doesn't go far enough into the North country for auroral losses to hit us.

```
                    S N A P m a x    from    W A 3 Z K Z
      SIGNALS, NOISE and PROPAGATION   - by -   Crawford MacKeand

F          Frequency (Openings & Analysis)  14.05   MHz
D    T     Date,   Time GMT                 10 Mar 2000        08:00
S   (N)    Solar Flux and A Index           150                15
P          Power & Required Bandwidth       100 W              250   Hz
           Receive Noise Level              Suburban           Zone   VK6
A          Antennas, Local & Remote         3 El Beam          3 El Beam

L          Local Site Name                Test Circuit to Remote #1
           Local Site Location           -32      South   -116.5    East
R   (C)    Remote Site #1 Name            Delaware (W3)
           Remote Site Location           39      North    75.5     West

Mode       Single calculation shown for analysis purposes.
B          Bandwidth and signal/noise thresholds for CW modulation.
E          Enter set-up, default & information menu


  Site        Name       Angle    F(km)      Latitude    Longitude    Az El
                                          (Geog.) (Mag.)
  Remote  Delaware                         39°      50.9°      75.5°    302°
  RP1  Reflection Rem1,  Low     365      45.2°    56.4°      92.4°    4.7°
  RP3  Reflection Rem1, High     331      44.5°    55.8°      89.8°    7.6°
 (RP5)  .. Mid Point ..          304      28.3°    23.9°     -178°       F
  RP4  Reflection Loc2, High     341     -24.1°   -35.6°    -126.7°    7.6°
  RP2  Reflection Loc2,  Low     328     -22.7°   -34.1°    -128.3°    4.7°
  Local   Test Circuit to Remote #1      -32°     -44°      -116.5°    51°

  Date         10 Mar 2000     Time    08:00 GMT     Distance  18680   km
  Frequency   14.05 MHz        10 cm Flux, SSN & A Index  150    105    15
  Number of hops,  low angle path -   6      high angle path -   7

  Noise, atmospheric & galactic        35.3    dBkTo         25.5 dBkTo
  Manmade noise                        39.4    dBkTo         (SU)
  Total noise power at 1Hz & BW        9.4     dBkTob       -109 dBm
  Digital Speed & Multipath Delay      8       Baud          .5 ms (min.)
```

Figure 12.2 Circuit Definitions for the Proposed Route W3 - VK6

Looking now at the daily picture in Figure 12.3, below, which shows signal-noise ratio in 6 dB S-units, (the screen picture can take advantage of color and so shows more information), the best time to try this route appears to be around 1600 to 1700 UTC (GMT), with a MUF of 18 to 19 MHz and an LUF of 12 to 13 MHz for our specified equipment. So now let's take a look at the details, both for 0800 GMT and then for 1630 UTC, about mid-day in Delaware and mid-night in Perth.

```
--------------------------------------------------------------------------*
          SIGNAL TO NOISE (S-Units)         - SNAPmax -   by WA3ZKZ

 10 Mar 2000  10.7 cm Solar Flux  150    A Index  15     CW modulation
 Local Site  Test Circuit #1 W.A. (VK6)  Latitude -32    Longitude -116.5
 Remote Site Delaware                    Latitude  39    Longitude  75.5
 Local Beam Heading 51 degrees      Remote Beam Heading 302 degrees
             --- Great Circle Distance 18680  Km.  ---
--------------------------------------------------------------------------*

GMT                                                                 MHz
TIME   0    5    10   15   20   25   30   35   40   45   DAY  LUF     MUF
----   !----!----!----!----!----!----!----!----!----!    *   -----------
  0    |               011**--                          **.  16      18.9
  1    |              .0**-                             **.  16      16.6
  2    |              *--                               **.          14.4
  3    |            *-                                  **.   -      12.2
  4    |          *-                                    **.   -      10.4
  5    |          *-                                    **.   -      10.8
  6    |           *-                                   **.          11.4
  7    |          .**-                                  *..          11.9
  8    |         .1**-                                  *..  11      11.8
  9    |         01*--                                  *..  10      11.7
 10    |         .**-                                   *..          11.0
 11    |          *-                                    ...   -      10.4
 12    |          *-                                    ..*   -      10.6
 13    |            0*-                                 ..*  13      13.1
 14    |            012**-                              ..*  13      15.8
 15    |           .012334**--                          ..*  13      18.6
 16    |           01233445**--                         ..*  12      19.2
 17    |          .012344*--                            ..*  12      17.2
 18    |          0123**-                               ..*  11      14.8
 19    |         .023*-                                 .**  11      13.3
 20    |          .1**-                                 .**  12      12.8
 21    |            0122**--                            .**  13      16.9
 22    |              012233**--                        .**  15      20.2
 23    |               .01233**---                      ***  17      21.7
----   !----!----!----!----!----!----!----!----!----!   LMR  -----------
       0    5    10   15   20   25   30   35   40   45         FREQUENCY
```

Figure 12.3 Daily Overview, Direct Circuit from W3 to VK6

```
        Path Parameters for  4.7°  F Ray Elevation Angle

Frequency 14.05 MHz   Great-Circle Dist. 18680 km    Modulation CW

Date and Universal Time UTC (GMT)     10 Mar 2000      08:00
Max. usable freq. (F) & min. foF2     11.8 MHz         3.7 MHz
MUF upper, lower bounds (90%, 10%)    14.8 MHz         9.8 MHz
G.C. dist. per hop & slant distance   3113 km          19653 km
Avg. F2 height                        365 km
Hop maximum and F-layer skip          4068 km          f over MUF
Distance and absorption losses        91.9 dB          18.4 dB
Land & oceanic reflection losses      1.1 dB           .4 dB
Auroral & over-MUF losses             0 dB             4.5 dB
Noise Power & circuit prob.          -109 dBm          16 %
Antenna gains, local & remote         5.5 dBi          8.5 dBi
Transmitted power & recvd signal      50 dBm          -93.3 dBm
Signal strength        S6             1.4 dBµV/m       1.17 µV/m
Signal-noise, actual & Mid. reqd.     15.7 dB          7 dB
```

Figure 12.4 Analysis of Proposed Circuit, W3 - VK6, 0800 GMT, 14.05 MHz

```
        Path Parameters for  4.6°  F Ray Elevation Angle

Frequency 14.05 MHz   Great-Circle Dist. 18680 km    Modulation CW

Date and Universal Time UTC (GMT)     10 Mar 2000      17:30
Max. usable freq. (F) & min. foF2     16 MHz           5.1 MHz
MUF upper, lower bounds (90%, 10%)    20.2 MHz         13.2 MHz
G.C. dist. per hop & slant distance   3113 km          19450 km
Avg. F2 height                        312 km
Hop maximum and F-layer skip          4053 km          2425 km
Distance and absorption losses        92.1 dB          23.8 dB
Land & oceanic reflection losses      1.1 dB           .4 dB
Auroral & over-MUF losses             0 dB             0 dB
Noise Power & circuit prob.          -110 dBm          84 %
Antenna gains, local & remote         5.4 dBi          8.3 dBi
Transmitted power & recvd signal      50 dBm          -94.6 dBm
Signal strength        S5             .2 dBµV/m        1.02 µV/m
Signal-noise, actual & Mid. reqd.     15.4 dB          7 dB
```

Figure 12.5 Analysis of Alternative Circuit, W3 - VK6, 1730 GMT, 14.05 MHz

This looks much better, and if transmission conditions are the only consideration, Figure 12.5 shows that we have a pretty good chance of making the QSO. But of course, there are always competing signals, which may make the contact difficult or even impossible, and the question of working hours .... will anybody be there? Since VK6 is not far from the opposite side of the world, maybe we could also take a look at the long path, and this is shown in Figure 12.6.

```
*---------------------------------------------------------------------------*
          SIGNAL TO NOISE (S-Units)          - SNAPmax -  by WA3ZKZ

  10 Mar 2000  10.7 cm Solar Flux  150    A Index  15      CW modulation
  Local Site   Test Circuit #1 W.A. (VK6)  Latitude -32     Longitude -116.5
  Remote Site Delaware                    Latitude  39     Longitude  75.5
  Local Beam Heading 231 degrees      Remote Beam Heading 122 degrees
             Long Path Great Circle Distance 21349  Km.
*---------------------------------------------------------------------------*

GMT                                                                MHz
TIME  0    5   10   15   20   25   30   35   40   45  DAY  LUF     MUF
----  !----!----!----!----!----!----!----!----!----!   *   -----------
0     |            .01**--                             *..  15   16.8
1     |             .*-                                *..       14.0
2     |           **-                                  *..   -   11.9
3     |            *-                                  *..   -   12.2
4     |          .**-                                  *..       12.8
5     |          .0*-                                  *..  13   13.2
6     |          .0*-                                  *..  13   13.2
7     |           .*-                                  **.       13.1
8     |            *-                                  **.   -   12.3
9     |           *--                                  **.   -   11.7
10    |            *-                                  **.   -   12.3
11    |               **-                              .*.   -   15.7
12    |                 .0**-                          .**  19   19.2
13    |                .01**--                         .**  18   19.9
14    |               .01**--                          .**  17   18.7
15    |               01**--                           .**  16   17.7
16    |              .11*--                            .**  16   17.0
17    |            .012*--                             .**  14   16.3
18    |           012**-                               .**  12   15.0
19    |          012*--                                ..*  11   13.5
20    |         .1**-                                  ..*  11   11.8
21    |         .0*--                                  ..*  11   11.8
22    |           **-                                  ..*       11.9
23    |             .0**--                             *.*  15   15.9
----  !----!----!----!----!----!----!----!----!----!  LMR  -----------
      0    5   10   15   20   25   30   35   40   45        FREQUENCY
```

Figure 12.6 Daily Overview, Long Path Circuit from W3 to VK6

Clearly not as good as the direct route, although some possibilities do show up, and if we saw a bad interference problem from West Coast stations then it might well be a better way to go. Although generally poorer, maybe it's not too bad at 1600 GMT; most likely quite workable with higher power, or better antennas, or a more resilient modulation scheme if we don't have tip-top ears at both ends!

```
*---------------------------------------------------------------------------*
                 SIGNAL TO NOISE (S-Units)          - SNAPmax -   by WA3ZKZ

 23 Mar 2001  10.7 cm Solar Flux  200     A Index  5        CW modulation
 Local Site  Test Circuit #2 Kansas (W0) Latitude  40      Longitude  100
 Remote Site Sweden                      Latitude  59.4    Longitude -18
 Local Beam Heading 29 degrees      Remote Beam Heading 313 degrees
              --- Great Circle Distance 7591  Km.  ---
*---------------------------------------------------------------------------*

 GMT                                                            MHz
 TIME  0    5   10   15   20   25   30   35   40   45  DAY  LUF    MUF
 ----  !----!----!----!----!----!----!----!----!----!   *   -----------
  0    |        .0122*-                                *..  10    13.2
  1    |         .011**-                               ...  11    13.8
  2    |           .0**-                               ...  13    13.7
  3    |          | **-                                ...        13.0
  4    |         |                                     ...   -    11.9
  5    |         |                                     ..*   -    11.3
  6    |         |                                     ..*   -    12.0
  7    |          |                                    ..*   -    12.5
  8    |          |                                    ..*   -    12.5
  9    |         |                                     ..*   -    12.0
  10   |         |                                     .**   -    11.2
  11   |        |                                      .**   -    11.0
  12   |         |                                     .**   -    12.3
  13   |           |                                   ***   -    14.4
  14   |             |  **-                            ***        16.5
  15   |               .01**--                         ***  17    18.5
  16   |             .011233**--                       ***  15    20.3
  17   |           ..011233444*---                     ***  14    22.0
  18   |          .0112233444***--                     **.  12    21.9
  19   |         01122334445**--                       **.  10    20.4
  20   |        0122233445**--                         **.   9    18.7
  21   |       .122333445*--                           **.   9    17.1
  22   |       01223344**-                             **.   8    15.6
  23   |       0122333*-                               *..   8    14.1
 ----  !----!----!----!----!----!----!----!----!----!  LMR  -----------
       0    5   10   15   20   25   30   35   40   45        FREQUENCY
```

Figure 12.7 Daily overview, relatively undisturbed ionosphere. Kansas - Sweden

Central USA to Northern Europe W0 - SM

A rather different picture is presented by a route from the State of Kansas to Sweden. This enters a part of the world where auroral influence is quite likely in the event of an ionospheric storm, so we will enquire what might happen under

```
*-------------------------------------------------------------------------------*
            SIGNAL TO NOISE (S-Units)           - SNAPmax -   by WA3ZKZ

    23 Mar 2001  10.7 cm Solar Flux  200    A Index  30      CW modulation
    Local Site  Test Circuit #2 Kansas (W0) Latitude  40     Longitude  100
    Remote Site Sweden                      Latitude  59.4   Longitude -18
    Local Beam Heading 29 degrees      Remote Beam Heading 313 degrees
                 --- Great Circle Distance 7591  Km.  ---
*-------------------------------------------------------------------------------*

    GMT                                                                MHz
    TIME  0    5    10   15   20   25   30   35   40   45   DAY  LUF    MUF
    ----  !----!----!----!----!----!----!----!----!----!     *   -----------
     0    |         | *-                                    *..   -    12.0
     1    |          |                                      ...   -    12.5
     2    |          |                                      ...   -    12.4
     3    |         |                                       ...   -    11.7
     4    |        |                                        ...   -    10.8
     5    |        |                                        ..*   -    10.2
     6    |        |                                        ..*   -    10.9
     7    |         |                                       ..*   -    11.3
     8    |         |  < FOT -- Optimum Working Frequency    ..*   -    11.3
     9    |        |         [Unavailable in this period]   ..*   -    10.9
    10    |        |                                        .**   -    10.1
    11    |       |                                         .**   -    10.0
    12    |        |                                        .**   -    11.1
    13    |          |                                      ***   -    13.0
    14    |            |                                    ***   -    15.0
    15    |             |     -                             ***   -    16.8
    16    |               |.0**--                           ***   18   18.4
    17    |             .0112**--                           ***   16   20.0
    18    |            .011233**--                          **.   14   19.8
    19    |         .0112233**--                            **.   12   18.5
    20    |       .0112233**--                              **.   10   16.9
    21    |       .012233**-                                **.   10   15.5
    22    |        01122*-                                  **.   10   14.1
    23    |         .0**-                                   *..   12   12.7
    ----  !----!----!----!----!----!----!----!----!----!    LMR  -----------
          0    5    10   15   20   25   30   35   40   45          FREQUENCY
```

Figure 12.8 Daily overview, somewhat disturbed ionosphere. Kansas - Sweden

disturbed conditions. We will select a solar flux of 200 and initially a relatively undisturbed A index of 5. Great circle distance is 7591 km., a moderate length path. If the A index is 30 the changes in probable circuit quality are quite clear!

Local Ground Wave Circuit

Local H.F. circuits are relatively unimportant in an overall picture, but may be extraordinarily important in an emergency situation, or highly desirable personally to complete a log for a *Five-Band Worked All States* award. The example below is for 100 watts of single sideband A.M. into quarter wave vertical antennas (much less probable at lower frequencies where an 80 meter installation would be 66 feet tall!). Medium conductivity ground is assumed; effective ranges would be much shorter in the arid Southwestern USA, but much longer between small Caribbean islands where useful paths would often be mainly over salt-water.

```
                      S N A P m a x   -by-   W A 3 Z K Z
   SIGNALS, NOISE and PROPAGATION  - by -  Crawford MacKeand    (V.501)

FREQUENCY    GROUND WAVE PROPAGATION    CIRCUIT QUALITY  Good  Mid. Poor
   MHz       Power  100 W    Antennas, Remote Vertical  Local Vertical
    2       GGGGGGGMMMMMPPPPPPPP...................
    4       GGGGGGMMMMMPPPPPPPP..................
    6       GGGGGGMMMMPPPPPPP...............
    8       GGGGGMMMMPPPPPP...........
   10       GGGGMMMPPPPPP.........
   12       GGGMMMPPPPP.......
   14       GGGMMPPPP.......
   16       GGGMMPPP......
   18       GGMMPPPP.....
   20       GGMMPPP.....
   22       GGMMPPP....
   24       GGMPPP....
   26       GGMPPP....
   28       GGMPP....
  Signal    ----------------------- Distance (Km) ---------------------
   SSB      0       20       40       60       80       100       120
```

Figure 12.9 Circuit Quality (Poor, Medium, Good) for Ground Wave Propagation

ACKNOWLEDGEMENTS

My primary education in radio propagation owes whatever depth it may have achieved to the classic studies of Jordan, Davies and Rawer. My deep gratitude to these three authors and also to all those consulted and detailed in the Bibliography. To those that I may have misconstrued or misinterpreted (and surely there are at least a few), my apologies. Data sources for several derived diagrams are acknowledged in the text, and I am grateful also to those authors. I also wish to acknowledge the many sources of information used in preparing this book, but especially the Morris Library of the University of Delaware, Newark Delaware. Last but not least, my thanks and 73 to many fellow radio amateurs who have contributed to those useful signals which may be detected amid my personal background noise level of experience.

Chapter X

Much of this work was originally presented as *Wide Band Signals from Poldhu?* by J.C.B. MacKeand, and M.A. Cross of the Historical Electronics Museum, Baltimore, Md., at the 1995 *International Conference on 100 Years of Radio*. The conference was organized by the Science, Education and Technology Division of the Institution of Electrical Engineers in association with the British Vintage Wireless Society and the International Union of Radio Science, marking the 100th. anniversary of Marconi's first radio demonstrations in England in 1895. It was held at the IEE, Savoy Place, London on 5 Sept. 1995 and papers were published as IEE Conference Publication Nr. 411. Michael Cross has continued since that time to provide very pertinent and greatly valued input, analysis and critique, and a particularly stimulating result of the IEE paper has been a considerable ongoing correspondence with Dr. Desmond Thackeray, of Byfleet, Surrey, England. He has challenged me with a variety of worthwhile insights, not least in our many digressions from any resemblance to a straight and narrow theme.

ACKNOWLEDGEMENTS

Chapter XI

The extensive data research to support the study of the Amelia Earhart flight and equipment was carried out almost entirely by David Dunlap, formerly a military pilot of extensive and varied experience, and now an airline pilot; without his extraordinary efforts the work would not have been started, nor could it have been finished. We are both much indebted to Walter Grosselfinger W2ATQ for very necessary conversation on the Western Electric aircraft equipment, and to many amateurs for leads to the RCA AR-60 receiver being restored by Bob Haworth W2PUA. The Coast Guard Historian, National Archives, Western Electric Archives, American Radio Relay League, Smithsonian Institution, Historical Electronics Museum in Baltimore, and many others have also provided essential leads and valuable help.

Investigation of this fascinating area of aeronautical and radio history has continued beyond the scope of this chapter, and in this matter David Dunlap and the author have both greatly enjoyed and particularly valued Capt. Elgen Long's strong interest and supportive comment. We would also like to acknowledge the interest of our friends at Nauticos Inc., Baltimore Md., and of members of Collins Amateur Radio Club, including Rod Blocksome K0DAS, Roger Hatcher WB9OMY, and Tom Vinson NY0V, with whom we have had stimulating e-mail discussions on several aspects of the problem.

BIBLIOGRAPHY AND REFERENCES

PROFESSIONAL TEXTS AND PAPERS

Aiya, S.V.C., *Atmospheric Noise Interference to Short-Wave Broadcasting*, Proc. IRE, March 1958 p580-589

Al'pert, Y.L., *Radio Wave Propagation and the Ionosphere*, Consultants Bureau NY 1963 (trans. from Russian)

Barghausen A.F., Finney J.W., Proctor L.L. & Schultz L.D., *Predicting Long-Term Operational Parameters of High-Frequency Sky-Wave Telecommunication Systems*, ESSA Tech. Report ERL 110-ITS 78 (NTIS Document N70-24144 Springfield VA., 1969)

Braun G., *Planning & Engineering of Short Wave Links*, Siemens/Wiley, London 1986

Bremmer H., *Terrestrial Radio Waves*, Elsevier, New York, Amsterdam 1949

David P. & Voge J., *Propagation of Waves*, Pergamon Press, Oxford, 1969

David P. & Voge J., *Propagation des Ondes*, Eyrolles Editeur, Paris, 1966. (in French, see 1969 trans.)

Davies K., *Ionospheric Radio Propagation*, Dover, NY 1966 (originally published as National Bureau of Standards Monograph 80, 1965)

Davies K., *Ionospheric Radio Waves*, Blaisdell/Ginn, Waltham MA. 1969

Davies K., *Ionospheric Radio*, Peter Peregrinus Ltd. (IEE) London 1990

Freeman, R.L., *Radio System Design for Telecommunications*, Wiley, NY 1997

Fricker R., *Microcomputer Program for the Critical Frequency and Height of the F Layer of the Ionosphere*, Fourth IEE Conf. on Antennas & Propagation, IEE 1985 p. 546-550

George P.L. & Bradley P.A., *A new method of predicting the ionospheric absorption of high frequency waves at oblique incidence*, Telecommunication Journal, 41/5 1974 p307-312

Gustafson W.E., Chase W.M., *Shipboard HF Receiving Antenna System; Design Criteria*. Naval Electronics Lab Ctr, San Diego, CA. AD872808 2 Jun 1970

BIBLIOGRAPHY AND REFERENCES

Goodman J.M., *HF Communications, Science and Technology*, Van Nostrand Reinhold, New York 1992.

Hall M.P.M & Barclay L.W. (eds.), *Radiowave Propagation*, Peter Peregrinus Ltd. (IEE) London, 1989

Hortenbach K.J. & Rogler F., *On the Propagation of Short Waves over Long Distances -- Predictions and Observations*, Telecommunication Journal 46/6 1979 p320-329

Jones, T.B. (ed.) *Oblique Ionospheric Radiowave Propagation*, AGARD Conf. Proceedings (13), Technivision Services, Slough, England 1969

Jordan E.C., *Electromagnetic Waves & Radiating Systems*, Prentice-Hall, NY 1950

Kessler E. (ed.) *Thunderstorm Morphology and Dynamics*, Universty of Oklahoma Press, Norman OK and London, 1986

Lovell B., and Clegg, J.A., *Radio Astronomy*, Chapman & Hall, London, 1952

McNamara L.F., *The Ionosphere: Communications, Surveillance and Direction Finding*, Krieger Publ., Malabar FL 1991

Maeda K.I. & Sato T., *The F-Region during Magnetic Storms*, Proc. I.R.E., Feb. 1959 p232-239

Picquenard A., *Wave Propagation*, (Philips Library), MacMillan, London 1974

Rawer K., *Calculation of Sky-Wave Field Strength*, Wireless Engineer, London, Nov. 1952 p287-300

Rotkiewicz, W. (ed.), *Electromagnetic Compatibility in Radio Engineering*, Elsevier, Amsterdam, Oxford, New York, 1982

Rush C.M., *Ionospheric Radio Propagation Model and Predictions*, IEEE Trans. on Antennas & Propagation AP-34 (9) Sept 86 p1163-70

Salaman R.K., *A New Ionospheric Reduction Factor (Multipath)*, IRE Trans. Commun. Syst. CS-10, 1962

Saxton, J.A. (ed.) *Advances in Radio Research*, V2, Academic Press, NYC 1964

Skomal, E.N., *Man-Made Radio Noise*, Van Nostrand Reinhold Co., NYC 1978

Terman, F.E., *Radio Engineering*, McGraw-Hill NY 1937

Terman, F.E., *Radio Engineers' Handbook*, Mc-Graw-Hill, NY 1943

Tremellen, K.W. and Cox, J.W., *The Influence of Wave Propagation on Planning of Short Wave Communication*, Proc. IEE, V.94, Pt. IIIa, 1947, p200-213

URSI, *The measurement of characteristics of terrestrial radio noise*. URSI Special Report No. 7, Elsevier, Amsterdam 1962

Volland, H. (ed.), *Handbook of Atmospherics*, Vol. 1, CRC Press, Boca Raton, Fla. 1982

Watt, A.D., *VLF Radio Engineering*, Pergamon Press, London 1967

ITU (CCIR), NBS (CRPL), & JTAC (IEEE) REFERENCE HANDBOOKS

Compendium of Technical Regulations, CCIF, CCIT, CCIR of the International Telecommunications Union, ["The Red Book"], Automatic Telephone & Electric Ltd., Liverpool, 1956

Interim Method for Estimating Sky Wave Field Strength and Transmission Loss at Frequencies between the Approximate Limits of 2 and 30 MHz. CCIR Report 252-2, New Delhi 1970

Man-Made Noise, CCIR Report 258-4, Dubrovnik 1986

Radio Noise, ITU-R Recommendation P.372-6, International Telecommunication Union, Geneva, 1993

World Distribution and Characteristics of Atmospheric Radio Noise, International Radio Consultative Committee (CCIR), International Telecommunications Union, Geneva, Report 322 1964 (NTIS Document PB 255 060, US Dept. of Commerce, Springfield VA)

Bandwidth, Signal-to-Noise Ratio and Fading Allowances in Complete Systems, CCIR - Recommendation 339-5 Kyoto 1978

Ground Wave Propagation Curves for Frequencies between 10 KHz and 30 MHz, CCIR - Recommendation 368-5 Los Angeles 1986

Ground Wave Propagation Curves for Frequencies between 10 KHz and 10 MHz, CCIR - Recommendation 368-2 Geneva 1974

Handbook for CRPL Ionospheric Predictions based on Numerical Methods of Mapping, Ostrow, S.M., (NBS Handbook 90), U.S. Dept. of Commerce, National Bureau of Standards, Washington DC 1962

Radio Spectrum Conservation, JTAC (Joint Technical Advisory Committee), IRE/RTMA, McGraw-Hill, New York 1952

Radio Spectrum Utilization, JTAC (Joint Technical Advisory Committee), IEEE/EIA, IEEE, New York 1964

BIBLIOGRAPHY AND REFERENCES

AMATEUR RADIO SOURCES

Bennington T.W., *Short-Wave Radio and the Ionosphere*, Iliffe, London 1950

Brown R.R. NM7M, *The Little Pistol's Guide to HF Propagation*, Worldradio Books, Sacramento CA 1996

Jacobs G. W3ASK, Cohen T.J. N4XX, & Rose R.B. K6GKU, *The New Shortwave Propagation Handbook*, CQ Communications Inc. Hicksville NY 1995 (extensive update of 1982 publication.)

Jacobs G. W3ASK and Cohen T.J. N4XX, *The Shortwave Propagation Handbook*, CQ Publishing, Inc., Hicksville NY 1982

Lewallen R.W. W7EL, *ELNEC*, (Antenna Program) Beaverton OR 1991 (based on MININEC, Logan, J.C. & Rockaway R.W., Naval Ocean Systems Center, San Diego.)

MacKeand J.C.B. WA3ZKZ/VP8CMY, *Path Losses at HF*, Communications Quarterly, Summer 1997 p9-16, with corrected graph Fall 1997 p105

Moxon L.A., G6XN, *HF Antennas for All Locations*, Radio Society of Great Britain, London 1982

Newton C., G2FKZ, *Radio Auroras*, RSGB, London 1991

Rose R.B. K6GKU, MINIMUF: *A Simplified MUF-Prediction Program for Microcomputers*, QST, Dec. 1982, ARRL, Newington CT

Saveskie P.N., *Radio Propagation Handbook*, Tab Books, Blue Ridge Summit PA 1980

The ARRL Handbook, American Radio Relay League, Newington CT

The ARRL Antenna Handbook, American Radio Relay League, Newington CT

The Radio Communication Handbook, Radio Society of Great Britain, (RSGB) London 1968

The Teleprinter Handbook, Radio Society of Great Britain, London 1973

INDEX

INDEX